# DATA VERSUS DEMOCRACY

## HOW BIG DATA ALGORITHMS SHAPE OPINIONS AND ALTER THE COURSE OF HISTORY

---

*Kris Shaffer*

Apress®

***Data versus Democracy: How Big Data Algorithms Shape Opinions and Alter the Course of History***

Kris Shaffer
Colorado, USA

ISBN-13 (pbk): 978-1-4842-4539-2 ISBN-13 (electronic): 978-1-4842-4540-8
https://doi.org/10.1007/978-1-4842-4540-8



Managing Director, Apress Media LLC: Welmoed Spahr
Acquisitions Editor: Shiva Ramachandran
Development Editor: Laura Berendson
Coordinating Editor: Rita Fernando

Cover designed by eStudioCalamar

Distributed to the book trade worldwide by Springer Science+Business Media New York, 233 Spring Street, 6th Floor, New York, NY 10013. Phone 1-800-SPRINGER, fax (201) 348-4505, e-mail orders-ny@springer-sbm.com, or visit www.springeronline.com. Apress Media, LLC is a California LLC and the sole member (owner) is Springer Science + Business Media Finance Inc (SSBM Finance Inc). SSBM Finance Inc is a **Delaware** corporation.

For information on translations, please e-mail rights@apress.com, or visit http://www.apress.com/rights-permissions.

Apress titles may be purchased in bulk for academic, corporate, or promotional use. eBook versions and licenses are also available for most titles. For more information, reference our Print and eBook Bulk Sales web page at http://www.apress.com/bulk-sales.

Any source code or other supplementary material referenced by the author in this book is available to readers on GitHub via the book's product page, located at www.apress.com/9781484245392. For more detailed information, please visit http://www.apress.com/source-code.

*Blessed are the peacemakers.*

# Contents

# About the Author

**Kris Shaffer,** PhD, is a data scientist and Senior Computational Disinformation Analyst for New Knowledge. He coauthored "The Tactics and Tropes of the Internet Research Agency," a report prepared for the United States Senate Select Committee on Intelligence about Russian interference in the 2016 U.S. presidential election on social media. Kris has consulted for multiple U.S. government agencies, nonprofits, and universities on matters related to digital disinformation, data ethics, and digital pedagogy.

In a former (professional) life, Kris was an academic and digital humanist. He has taught courses in music theory and cognition, computer science, and digital studies at Yale University, the University of Colorado Boulder, the University of Mary Washington, and Charleston Southern University. He holds a PhD from Yale University.

# Acknowledgments

How do you write the acknowledgments section for a book like this? If researching and writing this book has taught me anything, it's that some things are best said in private and, if possible, in person. So in lieu of public acknowledgments, I have decided to give a personal, hand-written note to those who educated, inspired, or otherwise helped me on this project. And hopefully, that note will be delivered in person and accompanied by a drink or a meal.

However, there is one group of people too large to thank individually, but whose influence and motivation have been immeasurable. To all my students over the years, who have inspired me with their hard work, their brilliance, and their desire to make the world a better place, I thank you. May each of you make your own dent in the universe and nudge humanity, even a little bit, in the right direction.

# Introduction: From Scarcity to Abundance

## A Brief History of Information and the Propaganda Problem

As long as we've had information, we've had disinformation. As long as we've had advertising, we've had attempts at "psychographic profiling." Ever since the invention of the printing press, we've had concerns about the corrupting influence of mass media. But there are some things that are new in the past decade. Information is abundant in a way we couldn't conceive of just a decade or two ago, and the new science of recommendation engines—algorithmically selected content, based on personal data profiles—dominates the modern media landscape. In this introduction, we will clear away misconceptions and focus on the heart of the problem—the intersection of information abundance, human psychology, user data profiling, and media recommendation algorithms. This is where propaganda finds its way into modern society.

## The Lay of the Land

Have you ever used a search engine to find a stock photo? Maybe you needed a slick header for your blog post, some scenery for your family holiday letter, a background for an office party flyer. The results can be pretty good, especially if you're on a professional stock image site (and know how to choose your search terms).

But have you ever used a regular search engine to find images of something generic? Try it sometime. Search for images of *doctor*, then *nurse*. *Professor*, then *teacher*. What do you see?

Chances are you found some pretty stark stereotypes. White-haired professors, wearing tweed, lecturing in front of a chalkboard. Teachers also in front of chalkboards, smiling at their eager pupils. Doctors in white coats, deftly wielding their stethoscopes or making notes on their patients' charts. Nurses in blue scrubs, also masters of their charts and scopes. You get the picture.

Walk through a school, a university, a hospital, a general practitioner's office, though, and you'll see little that matches these images. Most schools and universities abandoned chalk long ago, in favor of dry erase boards and electronic projectors. And lecturing to seas of students in rows is increasingly rare, particularly with young students. And, by the way, all of these professionals tend to dress less formally, and certainly more diversely, than the subjects of these search result images.

Search engines don't give us reality. They give us the results we *expect* to see. Using a combination of human programming, data from user interaction, and an ever-repeating feedback loop of the two, the results of these searches gradually become more like the generalizations in our minds. The stereotypes we hold in our minds determine what we click on, and those clicks form the raw data used by the search engine's algorithms. These algorithms, in turn, form generalizations of expected user behavior, based on our collective clicks, and serve up the results we're most likely to click on, based on that data. We perceive, we generalize, we search, we click, the machine perceives (the clicks), the machine generalizes, the machine returns results.

It doesn't end there. Those search results get *used* all over the web and in print (isn't that why we were searching in the first place?). Those images become part of the backdrop of our view of the world and further fuel the generalizations formed by our mind. Thus forms an endless loop: human perception → human generalization → human behavior → machine perception → machine generalization → machine behavior → human perception → human behavior ... And in each turn through the feedback loop, the stereotype gets more stereotypical. Reality is lost. Though, because the stereotypes become part of our media landscape, in a very real sense, reality is also *formed*.

But did you notice something else strange about those image results?

How many of those doctors were men, and how many were women? What about the nurses? According to *The Wall Street Journal*, 32% of doctors in the United States in 2012 were women, and the proportion is rising.[1] Is that the percentage you saw in your search results? According to the National Center for Education Statistics, 49% of tenure-track university faculty and 57% of non-tenure-track faculty in the United States are women.[2] How did your *professor* search results compare?

[1] Josh Mitchell, "Women Notch Progress," *The Wall Street Journal*, published December 4, 2012, `www.wsj.com/articles/SB10001424127887323717004578159433220839020`.
[2] "Quick Take: Women in Academia," *Catalyst*, published October 20, 2017, `www.catalyst.org/knowledge/women-academia`.

Chances are your search results were more stereotype than reality. That's partly our brains' fault. Our brains make generalizations about what we perceive in the world, and those generalizations allow us to make predictions about the world that help us interact with it more efficiently. Cognitive scientists have also found that when we form generalizations—called *schemas*—we tend to define those schemas, in part, by contrast with other schemas. In other words, we emphasize their *differences*, often making them *more distinct* from each other in our minds than they are in reality. While this method of defining ideas and categories in our mind is usually helpful, it sometimes works against us by reinforcing the *bias* of our environment, including the (stereotype-ridden) media we encounter. And when the feedback loop of human generalizations, machine generalizations, and media representation goes online, that bias gets propagated and reinscribed at near light speed.

It's all connected—our media, our memory, our identity, our society. The way we interact with the world is directly influenced by the "mental map" we have of the world—what's real, what's not, and where it all belongs. And while that mental map is primarily formed in light of our experiences (with a little help from hundreds of thousands of years of human evolution), that map is increasingly influenced by the media that we consume.

I say "increasingly" not because something in our brains has changed, but because our media landscape has changed so drastically in the last century—even the last decade. We have moved from information scarcity to information abundance, from media desert to media ubiquity. And for most—though, importantly, not all—humans on this planet, our access to the information and media that exists has expanded just as rapidly. Our lived experiences are increasingly mediated through media.

## The Limits of Attention

But one important thing has *not* changed: the limits of the human body, including the brain. Sure, infant mortality[3] and life expectancy[4] have improved in most societies over the past century, and quality of life has improved for many as the result of scientific and humanistic advancement. But the human cognitive system—the interaction of the brain and the body, memory and the

[3] "Infant Mortality," World Health Organization Global Health Observatory (GHO) Data, accessed February 5, 2019, www.who.int/gho/child_health/mortality/neonatal_infant_text/en/.

[4] "Life Expectancy," World Health Organization Global Health Observatory (GHO) Data, accessed February 5, 2019, www.who.int/gho/mortality_burden_disease/life_tables/situation_trends_text/en/.

senses—took its more-or-less modern form tens of thousands of years ago.[5] The amount of information our brains can hold has not changed, nor has the limits of conscious *attention*.

And with all the media clamoring for our attention, that attention has become our most precious—if also our most overlooked—resource.

Let's step back to the world of search engines for a moment. They can be hacked. I don't mean a full-on security breach (though that's certainly possible). I mean they can be manipulated by highly motivated users. Just as what people click on determines (in part) the search result rankings, the terms that people search for determine (in part) the terms that pop up on autocomplete as you type. (For some fun, go to Google and search for your home country/state: "Why is [Colorado] so" and see how Google thinks you might want to complete that search.) If the searches typed into a search engine can determine the autocomplete terms, then a group of people willing to put the time in can search for the same thing over and over again until it dominates the autocomplete results, increasing the number of people who see it, are influenced by it, and click on it.

This very thing happened in 2016. Members of the so-called alt-right movement (right-wing extremists, often affiliated with groups espousing hateful views like white supremacy, anti-Semitism, and hyper-masculine antifeminism) successfully manipulated Google's autocomplete to suggest racist searches and pro-alt-right messages.[6] When this manipulation was brought to Google's attention, they responded with changes to the system. But no system is completely impervious to hacking.

Not even the human mind.

## Cognitive Hacking

Any system that draws conclusions based on statistics, in part or in whole, can be gamed by a manipulation of the statistics. That's how the human brain works—over time, the things we perceive are aggregated together into generalized schemas, which are constantly changing with new information. (The more established schemas change more slowly, of course.) By altering the statistical input of the brain, a "hacker" can impact the schemas our brain

---

[5] Though scientists still debate just how many tens of thousands. See Erin Wayman, "When Did the Human Mind Evolve to What It Is Today?," *Smithsonian Magazine*, published June 25, 2012, `www.smithsonianmag.com/science-nature/when-did-the-human-mind-evolve-to-what-it-is-today-140507905/`.

[6] Olivia Solon and Sam Levin, "How Google's Search Algorithm Spreads False Information with a Rightwing Bias," *The Guardian*, published December 16, 2016, `www.theguardian.com/technology/2016/dec/16/google-autocomplete-rightwing-bias-algorithm-political-propaganda`.

forms. This *cognitive hacking* impacts not only how we define things but also what we associate them with and how positively or negatively we react to them. And just as the alt-right Google hackers were a minority of Googlers who were still able to impact the platforms, even small groups of cognitive hackers can have a noticeable impact on our "mental map" (and the actions it influences) if they can manipulate a large enough portion of the media we consume around a particular issue.

Now before you start thinking "The Russians! The Russians!", it's worth noting that a lot of cognitive "hacking" is inadvertent. (We will get to Russian, state-sponsored influence operations later in this book, but they are but a part of a much bigger problem.) That is, the structure of the media platforms we use on a daily basis has an impact on the way we perceive the world.

For example, we know that users engage more with media that elicits strong emotions, particularly if that emotion is anger.[7] We also know that people are increasingly consuming news via social media platforms,[8] which deliver content based on an algorithm that seeks to maximize "engagement" (clicks, likes, favorites, shares, angry emoji, etc.), and attempts to maximize engagement often translate into maximizing strong emotions, especially anger. It's no surprise, then, that those who study politics and digital media observe an increased polarization around political issues—strong emotions, diminished nuance, and ideological positions clumped further from the center.

This natural draw of our attention to things that elicit strong emotions is exacerbated by the *competition* for our attention, as the amount of data, and data producers, is growing at mind-boggling rates. In 2013, scientists estimated that 90% of the data in the entire world had been created in the previous two years alone.[9] And by 2015, Facebook alone had more users than *the entire internet* did in 2008.[10] Add to that the proliferation of TV channels in many households, and the portable prevalence of other digital screens, and we have a society marinated in visual and aural media that our evolutionary history simply did not prepare us for. As every bit of commercial media competes for our attention, each one does so by trying to outdo the other.

---

[7] Bryan Gardiner, "You'll Be Outraged at How Easy It Was to Get You to Click on This Headline," *Wired*, published December 18, 2015, `www.wired.com/2015/12/psychology-of-clickbait/`.

[8] Jordan Crook, "62 Percent of U.S. Adults Get Their News from Social Media, Says Report," *Techcrunch*, published May 26, 2016, `https://techcrunch.com/2016/05/26/most-people-get-their-news-from-social-media-says-report/`.

[9] SINTEF, "Big Data, for Better or Worse: 90% of World's Data Generated over Last Two Years," *Science Daily*, published May 22, 2013, `www.sciencedaily.com/releases/2013/05/130522085217.htm`.

[10] Paul Mason, *Postcapitalism: A Guide to Our Future*, New York: Farrar, Straus and Giroux (2015), p. 7.

The result is a digital arms race of cognitive manipulation, amplifying the already dangerous, polarizing effects of attention-grabbing media.

This polarization is the result of more than just emotion, though. People tend to engage with things that reinforce their existing point of view, what scientists call *confirmation bias*—if I agree with it already, there must be something good about it. This includes things that seem to confirm its accuracy—after all, it feels good to discover new information that proves you were right all along!—or to confirm the inaccuracy of the people who hold a different position. We have an evolved (and important!) predisposition to dislike, and avoid, circumstances that surprise us with our own faults. In a past era of regular predator-prey interactions, such situations were often deadly. But when it comes to political debate online, that evolved predisposition encourages us to avoid situations where we can grow intellectually, ethically, and morally, both as individuals and as a society. By avoiding nuance and succumbing to confirmation bias, we again move further from the center—and from those we disagree with. We build our own personal echo chambers, all the while believing that we are basing opinions on a *wider* base of information, thanks to the wonders of the world-wide web. Under this delusion, and without intervention, society becomes more fragmented, and it is harder for people from differing ideologies to find common ground.

Though we live in the most information-rich, most connected era in human history, the way our minds *naturally* deal with such information abundance means our generation is probably the most susceptible to propaganda in human history. But I stress *naturally*. Our genetic predispositions, hard won by our ancestors as they fought off threats from every side, are in many ways a tremendous gift. But in an era where most of us are *not* regularly threatened by predators, but by people who would seek to use media to manipulate our attention to influence our minds and our behavior, we are *naturally* equipped with the wrong tools for dealing with these threats. We have a great instinct for what to do when a predator surprises us from behind, or when we are low on calories and confronted with a bounty of food to eat. But we have no instinct for digital disinformation.

However, we *have* evolved the capacity to learn, to analyze, to reason, to communicate, to persuade with logic. These are not our *automatic* instincts. But if we direct our minds appropriately—if we master our own attention in the right ways at the right times—we can resist this manipulation and build a better society. But it takes work. And before that, it takes an understanding of the problem.

That's what we'll explore in this book.

## Where We're Headed

We'll begin by unpacking the "propaganda problem." Why is our conscious attention so limited? How limited is it? And what are the implications for how we engage information in the digital age? We'll also explore the algorithmic basis of social media and other content recommendation systems (like Pandora, Netflix, and Amazon)—how they work technically, how they interact with the strengths and limitations of our cognitive system, and how they can be "hacked" to influence our thoughts and actions.

Then we'll explore several case studies in how social media–based operations have significantly impacted public consciousness—even public action. From the Ferguson protests, organized and proliferated on Twitter and Vine, which led to significant growth of the Black Lives Matter movement, to the coordinated harassment of female game designers and critics on Twitter known as GamerGate, to the influence operations—both domestic and international—at work in recent elections and political uprisings in Ukraine, Sweden, the United States, Turkey, Brazil, the Philippines, Myanmar, Mexico, and Colombia.

We live in an era different from any we've ever seen before. We have the tools, the information, and the interpersonal connections to accomplish amazing things. But like the discovery of fire or the harnessing of nuclear energy, our newfound digital capabilities also leave us vulnerable to human-made disaster, the likes of which we've only begun to dream up. This power can be used for good, used for evil, or treated as a toy. By too often making digital technology a mere plaything, we open ourselves to those who may use it for our undoing. But with a little education, a little regulation, and a lot of care, we can turn this digital technology back into a beacon of hope.

It is my aim in this book to bring that education, foster that care, and, yes, encourage that regulation. Because if we work together, and we get it right, we *can* do some amazing things.

So let's get started.

PART

I

# The Propaganda Problem

CHAPTER

1

# Pay Attention

## How Information Abundance Affects the Way We Consume Media

This chapter will explain the shift from an information economy to an attention economy and lay out the implications for how information is created, shared, and consumed on the internet. Having transitioned from a time of information scarcity to information abundance, information is no longer a sufficiently monetizable commodity to drive an economy. The focus of human attention as the monetizable commodity in limited supply gives content recommendation algorithms a pivotal place in our information landscape and our economy. This chapter lays out the general economic, cognitive, and technological backdrop for the emergence of those algorithms.

## How Taste Is Made

What's your favorite dessert?

Mine is probably a nice flourless chocolate cake. With raspberries, blackberries, maybe some ice cream, … and, of course, something like an espresso, Americano, or a full-bodied stout to drink—something that can cut through that richness and cleanse the palate so that each bite is as wonderful as the first.

I feel like I gained a few pounds just *writing* that paragraph. Why is it that everything that tastes so good is so bad for you?!


K. Shaffer, *Data versus Democracy*,
https://doi.org/10.1007/978-1-4842-4540-8_1

Or is it?

When I first studied evolutionary psychology in college, I remember learning that humans generally evolved preferences for things that are *good for us*.[1] Things that are safe and help us live at least to the age of reproduction, so we can pass on our genes. Humans who associate something with a positive emotion will do that thing more often. Things that we associate with negative emotions we tend to avoid. We seek out pleasure more than pain, and our ancestors who took pleasure in things that helped them survive and reproduce, well, they survived and reproduced. Those who took pleasure in things that put them in danger, well, less of those genes were passed on to future generations. The result is that the genes we twenty-first-century humans have inherited generally make us feel pleasure around things that are good for the species, but produce pain to warn us away from things that would jeopardize our collective survival.

Of course, this emotional preference, this *taste*, for things that are biologically and socially good for the species also applies to food. Foods that are good for us tend to give us pleasure, while things that are bad for us tend to taste bad, or trigger our gag reflex.

Wait a second. That seems backward. Didn't I just lament about how everything that tastes *good* is actually *bad* for us?!

There's another piece to this evolutionary puzzle. Those evolved preferences, called "adaptations," helped our ancestors adapt to their environment. To *their* environment. Let's go back to that flourless chocolate cake. What are the ingredients? Cocoa (of course), sugar, salt, butter, and lots and lots of eggs. Nutritionally speaking, there's not a lot of vitamins, minerals, or fiber—things we hear a lot about today. Instead there are carbohydrates—"simple" sugars, to be precise—fat, and protein. We hear a lot about these things from modern nutritionists, too. They lead to overweight, obesity, heart disease, diabetes, kidney disease, even gum disease and tooth decay. These ingredients sit nearer the top of the food pyramid, things we should consume sparingly.

If we should consume them sparingly, then why did we evolve such a strong taste for them? For those of us in the affluent West, 2000 calories a day are easy to come by. But the genes that determine our culinary tastes *today* were not selected in an era of such abundance. Those genes have changed little since the Pleistocene period, tens to hundreds of thousands of years ago.[2] Our metaphorical palates were developed in a time before grocery stores, fast food restaurants, fine dining, potlucks, refrigerators, freezers, even a time

---

[1]Steven Pinker, *How the Mind Works*, New York: W. W. Norton & Company, Inc. (1997), pp. 524–25.

[2]For a detailed discussion, see Jerome H. Barkow, Leda Cosmides, and John Tooby (eds), *The Adapted Mind: Evolutionary Psychology and the Generation of Culture*, New York, NY: Oxford University Press (1992).

before agriculture. Long before humans built a tractor, a plow, a saddle, even a bow and arrow, our modern food preferences were being written into our genetic code.

Carbohydrates, fat, protein. While today we highlight the dangers of consuming them in excess, all of them are essential for our bodies to produce energy, build muscle, and keep in good working order. All of them are relatively easy to come by in a society with the technology to farm and hunt efficiently, not to mention the technology to preserve food and all its nutritional value for long stretches of time. But 150,000 years ago on the African savannah, our ancestors had no such technology. Those biological essentials—carbs, fat, protein—were *scarce*. Incredibly scarce. The human who took pleasure in consuming them would seek them out and, presumably, be more likely to find them and survive. And in an era when many humans lived on the brink of starvation, those who did not seek out those precious ingredients often did not survive to pass along their genes to us. It was the survivors who authored our genetic code, and it is their taste for what was scarce but essential aeons ago that governs our culinary predilections today.

And so we love things like flourless chocolate cake. Not because our ancestors adapted a taste for chocolate cake, but because it's the perfect combination of all the essential, but scarce, things they evolved a taste for.

I mentioned before that our genetic code hasn't changed much since the Pleistocene era. That's because humans living in an era of abundance don't get to write the genetic code! If most of humanity is no longer living on the cusp of starvation, there is no food-related natural selection going on. Death drives evolution, and genetic flaws get selected out, not the other way around. Take allergies as an example: my spouse and I have severe enough allergies that had we lived 100 years earlier, we both would have died from anaphylaxis before we reproduced. Thanks to modern medicine, we survived! And we both passed our allergy genes onto our children, leading to some severe medical situations in just the first few years of their lives. Once we're no longer on the brink of extinction, those *health* advantages aren't *evolutionary* advantages. And so, what was advantageous in our species' distant past still governs our preferences today.

## Supply and Demand: Why an Information Economy Is No Longer Sustainable

So I hear some of you asking, "What does all of this have to do with big data, 'fake news,' and propaganda?"

Everything.

Our ancestors' taste for food was determined in a time of food *scarcity*. The ingredients that were essential but scarce became the ingredients that were most preferred and sought out. However, in an era of *abundance*, those preferences don't fit, especially when the same ingredients that are essential in small to moderate amounts become harmful in large amounts. Our natural instincts for food have become dangerous in an era of relative abundance.

The same can be said of how we consume information. The way we deal with information is based on an evolutionary history—and an educational experience—in which information was *scarce* or at the very least *expensive*. Most of our cognitive system, which governs how our senses connect to our brain, evolved before *Homo sapiens* even existed.[3] The bulk of our genetic code was pretty well fixed long before the invention of writing, let alone the printing press, mass media, radio, television, or the algorithmic news feed.[4] Our instincts for determining reality, for parsing safety from danger, are not adapted to the media environment in which we live. We're so tuned for information scarcity that consuming modern media is like trying to drink from a firehose.

On top of that, most of us received an education that was designed during and for an internet-free world. The tools and techniques that help someone at a computer-free home, or even in the public library, judge the veracity of a claim they read are not fine-tuned for life on the internet. The truth is, simply put, that *most of us don't know how to fact-check on the internet*. We don't know how to do it, we don't know how to *make* ourselves do it. And when that happens on a social media platform where we can share everything we read with the tap of a finger, we compound the problem, as our uncritical *consumption* of media leads to uncritical *publishing* of media. Every time we read without our "crap detector" fully engaged, the more "crap" we inadvertently pass along. Our social media "stream" then doesn't just become a firehose, it becomes a firehose full of sewage. Finding something edifying becomes all the more difficult.

This transformation isn't just about evolution or education. It's also about the economy. We've all learned about the industrial revolution, when—at least in

[3]Most of the basic human cognitive functions are shared with other animal species, especially other mammals, and thus were likely possessed by our shared ancestors millions of years ago. But even some of humanity's more advanced cognitive capacities are shared with other primates, as discussed in fascinating detail in Douglas Fox, "How Human Smarts Evolved," *Sapiens*, published July 27, 2018, `www.sapiens.org/evolution/primate-intelligence/`.

[4]While scientists are still researching and debating when exactly humans developed the cognitive functions we know and rely on today, there is strong evidence that *Homo sapiens* possessed the high-level cognitive skills of spoken language and even music *at least* 50,000 years ago. See Steven Mithen, *The Singing Neanderthals: The Origins of Music, Language, Mind and Body*, London: The Orion Publishing Group, Ltd. (2005), pp. 260–65.

the West—the feudal, agrarian-based economy (and the accompanying social order) gave way to a market-based, industrial economy. Rule by land, blood, or divine right gave way to rule by money and labor. The aristocracy was replaced by merchants. Farms by factories (and, later, factory farms). And a whole new social order emerged. As economist Michael H. Goldhaber writes:

> Europe back then in the 15th century was still ruled pretty much on feudal lines, and the feudal lords took it for granted that the new world would be a space for more of a feudal economy, with dukes and counts and barons and earls ruling over serfs throughout the newly discovered continents. They did in fact begin to set up that system, but it was not what turned out to flourish in the new space. Instead, the capitalist, market-based industrial economy, then just starting out, found the new soil much more congenial. Eventually it grew so strong in North America that, when it re-crossed the ocean, it finally completed its move to dominance in Western Europe and then elsewhere in the world.[5]

In other words, there's a reason democratic revolution exploded among Europeans in North America before it ignited Europe itself. The lack of an established aristocracy and an emphasis on trade and material goods (brought home to Europe on the backs of slaves) made the New World a different kind of economic space—one where the mining and production of material goods, the ability to move it, and the ability to convert it into *capital* gave one power. This market-based capitalism, and with it the idea that anyone could better themselves in society through labor and intelligence, formed the basic of the new republic (if, at times, in theory more than in practice). Eventually it did reach Europe, as feudalism slowly gave way in many places to market capitalism.

But this isn't the only socioeconomic transition humanity has faced. Feudalism was not the first social system, nor is capitalism the last. (It's not even the only system around today!) And capitalism itself comes in many forms. Some have marked an important economic transition from a goods-and-services-based economy to an *information* economy. The idea is not that Western capitalism has gone away, but that the rules of supply and demand that once applied to materials (goods) and labor (services) now apply to information (data).

Think of it this way. When something is scarce, but desirable, we call it "precious." Gold, silver, fertile soil, fats, carbs, protein, etc. In a market-driven economy, if demand is high and supply low, people compete to trade

[5]Michael H. Goldhaber, "The Attention Economy and the Net," *First Monday* 2/4, published April 7, 1997, `http://journals.uic.edu/ojs/index.php/fm/article/view/519/440`.

for it, and in a capitalist economy, they trade money for it. Its cost goes up, a reflection of the value that society ascribes to it. If supply is high and demand is low, price goes down, a reflection of the relatively low value given to it by society.

In an industrial economy, the supply-and-demand equation that keeps things moving involves materials and labor—what can you get, what can you make with it, where can you deliver it? The people who control the raw materials and the labor rule the roost. But what happens when we get so good at making things that supply overpowers demand *even for the things we need most*, like food, clothes, or housing? This is what has happened in many developed countries, as mechanization and automation have made production so efficient as to drive supply sky high, and at the same time drastically reducing the need—and therefore the monetary value—of human labor. When those two things happen, no one has any money to spend, and when they do spend it, things are so cheap that no one *makes* any money. The economy grinds to a halt.

But, as Paul Mason has noted in his book *Postcapitalism: A Guide to Our Future*,[6] capitalism is rather resilient. As the supply-and-demand equation collapses in on itself in one economic sphere, another one often emerges to take its place. In the past few decades, as the cost of production (and, thus, the wages for labor) has generally decreased in the affluent West, *information* has emerged to take its place. In other words, it was no longer those who controlled materials and labor who ruled the roost, it was those who managed *information* who had the strongest hold on the economy—and the social order.

But before we could enter into a new, centuries-long era of info-capitalism, something else happened. Moore's Law happened. Named after engineer and Intel co-founder Gordon Moore, Moore's Law states that the number of transistors that fit in a given amount of space in a semiconductor circuit doubles roughly every two years.[7] In an era where the economy is ruled by supply and demand of information, that exponential growth of the ability to process information is a problem. Add to that the just as rapidly diminishing cost of *storing* information, and the absolutely trivial process of *copying* information, and the information economy was served a death warrant before it ever had a chance to go out on its own.

---

[6]Paul Mason, *Postcapitalism: A Guide to Our Future*, New York: Farrar, Straus and Giroux (2015).
[7]Gordon E. Moore, "Progress in Digital Integrated Electronics," Intel Corporation (1975), p. 3.

All the way back in 1984 at the first Hackers Conference, Stewart Brand famously said:

> On the one hand, information wants to be expensive, because it's so valuable. The right information in the right place just changes your life. On the other hand, information wants to be free, because the cost of getting it out is getting lower and lower all the time. So you have these two fighting against each other.[8]

Setting aside the problematic mantra that "information wants to be free" has become, this duality (which Cory Doctorow calls "quite a good Zen koan"[9]) is the fundamental problem of basing an economy on information. We can't. This is why so many publishers and news media companies have struggled to move from print and broadcast to digital. This is why we have paywalls, ad blockers, ad blocker blockers. You can't have a market capitalism that's based around a commodity that is cheap to produce, cheap to copy, and—now that we have the internet—cheap to access. Some economic voices, like the aforementioned Paul Mason, suggest that this means capitalism itself is dead, and that the information age will usher in a new, quasi-socialist era of *postcapitalism*. And maybe they're right.

Or maybe there's another commodity, and the resilient capitalism still has a little fight left in it.

When Goldhaber described the transition from feudalism to capitalism, he didn't stop there. He was actually writing about the emergence of a new economy. And not just that, but a new *kind* of economy.

> Contemporary economic ideas stem from that selfsame market-based industrialism, which was thoroughly different from the feudal, subsistence-farming-based economy that preceded it. We tend to think, as we are taught, that economic laws are timeless. That is plain wrong. Those laws hold true in particular periods and in a particular kind of space. The characteristic landscape of feudalism, dotted with small fields, walled villages, and castles, differs markedly from the industrial landscape of cities, smokestack factories and railroads, canals, or superhighways. The "landscape" of cyberspace exists only in our minds, perhaps, but even so it is where we are

---

[8]Cited in Doctorow, *Information Doesn't Want to Be Free: Laws for the Internet Age*, p. 94.
[9]Ibid.

> increasingly coming to live, and it looks nothing like either of those others. If cyberspace grows to encompass interactions between the billions of people now on the planet, those kinds of interaction will be utterly different from what prevailed for the last few centuries, or ever before.
>
> If you want to thrive in this new world, it behooves you not to mistake it for a place where the dukes and earls of today will naturally continue to prosper, but rather to learn to think in terms of the economy natural to it.[10]

What are these terms? Goldhaber, writing in 1997, just a few short years after the invention of the world-wide web (and, if I'm not mistaken, the year I received my first Juno email account), calls this new economy the *attention economy*. (Goldhaber wasn't the first to use the term, but he was one of the first to discuss in some detail what the advent of the web might mean for the socioeconomic structures of our post-industrial society.) In an attention economy, the supply-and-demand equation that governs economic growth does not concern the supply and demand of information, but the *supply and demand of human attention*. As Matthew Crawford says, "Attention is a resource—a person has only so much of it."[11] And as information becomes cheaper and easier to create, access, distribute, and copy, attention becomes increasingly scarce, at least from the perspective of those who deal in information.

As Goldhaber describes it, this isn't a mere shift from one kind of material for sale to another. It's a totally different kind of system. Where money moves in the *opposite* direction as goods and services (a modern version of trade), money moves in the *same* direction as attention. Yes, we're still buying things (and yes, content producers still buy advertising, which is kind of like buying attention), but increasingly we don't set out to buy what we want or need. We set out to be entertained, and the winners of the battle for our attention are the ones that receive our money. This means different kinds of transactions, different kinds of sales and marketing strategies, and different kinds of markets in which these transactions take place. (In industrial terms: is Facebook a media company, a social club, a public gathering space, or an advertising agency?)

---

[10] Goldhaber, "The Attention Economy and the Net."

[11] "Introduction, Attention as a Cultural Problem," *The World Beyond Your Head: On Becoming an Individual in an Age of Distraction* (hardcover) (1st ed.), Farrar, Straus and Giroux, p. 11.

## If You Don't Pay for the Product, You Are the Product: Attention as Commodity, Engagement as Currency

When it comes to disinformation and propaganda, there are a couple aspects of this economic transition that are critical to understand. First of all, the focus—the choke point, if you will—is our attention. The limits of our cognitive system are the bedrock of the economy that keeps the supply-and-demand equation in balance. If we are to be critical consumers, we have to get our heads around that, at least a little bit. Our attention is the primary "good" being traded online, on TV, in any form of media. In many ways, we play a *more* central economic role in these transactions, even if our money isn't involved. Our attention is the commodity, our "engagement" the currency—and, coincidentally, *this is why tracking our activity online is so important to companies that deal in information.* As long as the attention economy exists, Cambridge Analyticas will exist—companies that leverage data about customers, competitors, and even voters to provide tailored advertising to them with the express purpose of influencing, even manipulating, their behavior.[12]

Second, because our attention is the chief commodity of the new economy, the digital platforms that we use daily *are designed to manipulate and measure our attention.* As former Google employee and tech critic, Tristan Harris, put it, "A handful of people, working at a handful of technology companies, through their choices will steer what a billion people are thinking today."[13] Or put in perhaps a less dystopian tone, the tech that we use most is the tech designed to be so good that we keep coming back. But like food, alcohol, sex, driving a car, playing a sport, even good things can become dangerous when engaged carelessly or in excess.

But there's no escaping the fact that when platforms make their money off of advertising, they make more money when more people spend more time on those platforms. Here's the thing, though. While they make *some* money when you see an ad (called an *impression*), they make far more money when you *click* on the ad.[14] But that click takes you away from the platform, never to see

---

[12]For more on Cambridge Analytica in particular, see the groundbreaking journalism of *The Guardian* in "The Cambridge Analytica Files," www.theguardian.com/news/series/cambridge-analytica-files.

[13]Paul Lewis, "'Our Minds Can Be Hijacked': The Tech Insiders who Fear a Smartphone Dystopia," *The Guardian*, published October 6, 2017, www.theguardian.com/technology/2017/oct/05/smartphone-addiction-silicon-valley-dystopia.

[14]Greg McFarlane, "How Facebook, Twitter, Social Media Make Money From You," Investopedia, last updated March 21, 2014, www.investopedia.com/articles/investing/022315/high-cost-advertising-times-square.asp. See also, "How much It costs to advertise on Facebook," Facebook Business, accessed February 6, 2019, www.facebook.com/business/help/201828586525529.

another ad. Unless the platform is so good, so desirable, so worthy of your attention, that you come back soon after that trip elsewhere on the web.

In other words, the attention economy makes it good business sense to design platforms for addiction.[15] That's incredibly dangerous, both for individuals and for society. And when those platforms are increasingly the source of news for many individuals, as well as platforms often engaged when we are at our most relaxed, passive addiction becomes the way that many of us deal with the most important issues of the day. Harris says, "I don't know a more urgent problem than this. ... It's changing our democracy."[16]

Or put another way, this time with a more dystopian slant, our social media platforms are designed for propaganda.

## Algorithmic Recommendation: The Cause of, and Solution to, All of Life's Problems

Let's take a step back and get a general view on how this works. (We'll go into much more detail in Chapter 3.) Users want content. Whether that's text, music, TV, movies, all of that media demand is at its essence a demand for information. However, information is everywhere. As high as our demand for these various media is, supply far outstrips it. Both the amount of content and the ease with which we can access that content ensure that there's always plenty of free or inexpensive media we can access.

The problem, of course, is that content producers need to get paid, or they can't produce that content. (And while many content producers do it as a side hustle, our society would surely be impoverished if there were no longer any filmmakers, songwriters, novelists, or journalists who embarked on their craft as a full-time, long-term career.) The supply-and-demand equation is so in favor of the consumer that the average cost of creating media is rapidly approaching null.[17] The amazing thing, though, is that people are still spending money—a *lot* of it—on media. But as supply and access continue to increase, the chance of getting a piece of that pie is rapidly diminishing. So for a content producer to make money, they need to cut through the noise and be the ones to get our attention.

---

[15]Mike Allen, "Sean Parker unloads on Facebook: 'God only knows what it's doing to our children's brains'," *Axios*, published November 9, 2017, www.axios.com/sean-parker-unloads-on-facebook-god-only-knows-what-its-doing-to-our-childrens-brains-1513306792-f855e7b4-4e99-4d60-8d51-2775559c2671.html.

[16]Ibid.

[17]Cory Doctorow, *Information Doesn't Want to be Free: Laws for the Internet Age*, San Francisco: McSweeney's (2014), p. 55ff. See also Paul Mason, *Postcapitalism: A Guide to Our future*, New York: Farrar, Straus and Giroux (2015), p. 119.

Media consumers want content. Media producers need consumers' attention. More than that, media consumers want to cut through all the crap and find the good stuff—whether that's entertainment, education, or news. And media producers need to get their content in front of the *right* audience—the ones that might actually care enough to spend their hard-earned money on it.

Consumers want to find the right media. Creators want to find the right audience. And they're trying to find each other on the web. Kind of sounds like a dating site. In some ways, that's *exactly* what it is.

At its core, a dating site is a recommendation engine. By taking data about you and comparing it with data about other people, it recommends people with whom you might be compatible. The input data may come from an extended survey or a psychological personality test, or the input data could simply come from your behavior—hundreds of swipes, left and right. But ultimately, the process is the same. Take input data, run it through an algorithm, make recommendations. (And, in the better systems, collect data about how good the recommendation was, so the algorithm can be improved.)

When it comes to matching consumers with content, or media producers with audiences, the process is the same. Fine-tune an algorithm that matches a user with content, maximizing the compatibility so that consumers are happy with their choice, and producers get paid for their labor. The big difference with a dating site, though, is that if a dating site is successful, and you find the love of your life, you never need the dating site again. But if the media recommendation engine is successful, you'll not only find a great movie Friday night, you'll also come back for another one on Saturday. And remember, the *platforms* are also competing for our attention. Netflix, Hulu, Amazon, Pandora, Spotify, Twitter, Facebook, … the more they get us to come back, the more money *they* make, too, regardless of what media we consume.

This seems like a win-win-win situation. Platforms make recommendations; we find "personalized" entertainment choices that maximize our viewing/listening/reading pleasure; content producers have a way to find the *right* audience for their work, maximizing their income with minimal effort; and every time we give that track a thumbs up or thumbs down or rate that movie, the recommendations get better—for us and for the content producers. And as the recommendations get better, the platforms yield their own rewards. (When a media platform is also in the content creation business, it's a double win for them.)

But the system is not without its flaws. I've already discussed how it essentially primes us for addiction, prompting us to come back unconsciously, out of habit, rather than deliberately. But there are other problems. In order to make the best recommendations, platforms need lots of data about both us and the content to feed their algorithms. As platform competition gets fiercer, and computational resources cheaper, the amount of input data collected about us increases rapidly.

This data collection often happens without our knowledge. We may agree to the Terms of Service, but the details implied by those terms are not always transparent, and they sometimes change after we're already pretty well baked into the platform. This data collection also happens *when our guard is down*, when we're the least vigilant about what information we're providing to whom and to what ends. This isn't our bank statement, our mortgage agreement, our major term papers—the things we are careful and deliberate about. This is what we read first thing in the morning, before our first cup of coffee, sometimes before we get out of bed. This is what we listen to as we drive to work or go for a run. This is what we watch while we have a drink before bed. We're talking the music we listen to, the shows we watch, the news we read, what we say about it, the friends' pictures of their kids that we "like," the news stories that we respond to with a rage emoji, the pictures we spend the most time looking at, never mind the ads we click (and, in some cases, the credit card purchases that follow). All of that is logged, and much of it is used, even traded or sold, by the platforms, in order to serve up the content most likely to keep us on their platforms the longest and coming back the most often. This data collection leads to hacks, breaches, leaks, and, in some cases, targeted ads that know a bit too much about us. (Remember the case a few years back of the parent who found out their teenage daughter was pregnant because of the advertisements they received, before their daughter broke the news?[18])

There's another problem. For a platform to maximize our attention, and their data collection, they often strive to be one-stop shops. Amazon, Facebook, and Google (and before them, Yahoo!) are perhaps the best examples of this. The same "news feed" that gives us our news also gives us updates from our extended family, and for some serves as a professional development network. Then the ads encourage us to join groups where we can connect with people who share our religious convictions, or simply to buy a new pair of shoes or an upgraded smartphone. By maximizing our attention through such diverse content, the platform also *divides* our attention. They push us deeper into a state of "continuous partial attention," as Linda Stone calls it, where no one thing dominates our thinking.[19] This not only keeps us from thinking slowly

[18]Kashmir Hill, "How Target Figured Out a Teen Girl Was Pregnant Before Her Father Did," *Forbes*, published February 16, 2012, www.forbes.com/sites/kashmirhill/2012/02/16/how-target-figured-out-a-teen-girl-was-pregnant-before-her-father-did/.
[19]Linda Stone, "Beyond Simple Multi-Tasking: Continuous Partial Attention," Linda Stone (blog), November 30, 2009, https://lindastone.net/2009/11/30/beyond-simple-multi-tasking-continuous-partial-attention/.

and deeply about any one thing, but it causes us to constantly shift back and forth between different focuses. The resulting "attentional blink" (as some cognitive psychologists call it)[20] is a time of once again finding our bearings. And when we're constantly catching our breath, finding our place on the map, switching cognitive tasks, we find ourselves in what Stone calls "an artificial sense of constant crisis." And that's not just a psychological problem, that's a thinking problem. We can't slow down, we can't dive deep, we can't think critically. And the more time we spend on these platforms, the more time we spend in "attentional blink," in constant cognitive crisis.

When we find ourselves regularly in this psychological state *on the platforms where we find much of our news*, we are perfectly primed for propaganda. But just what is propaganda?

# Propaganda Defined

The word *propaganda* comes from the word *propagate*—to spread. In its oldest context, it simply refers to the spreading of a message, whether through word of mouth or through print media. It's similar both to *publishing* (making public) and to *evangelizing* (sharing good news), in that sense. But in more modern times, it's taken on a more sinister tone. In his classic text *Propaganda: The Formation of Men's Attitudes*, Jacques Ellul writes:

> Propaganda is a set of methods employed by an organized group that wants to bring about the active or passive participation in its actions of a mass of individuals, psychologically unified through psychological manipulation and incorporated in an organization.[21]

I find this definition helpful, but also insufficient for the digital age. The idea of an *organization* being the core agent, and expansion of that organization being the goal, only accounts for a small part of the propaganda activities we see online. The idea of being a card-carrying member of an organization has largely been supplanted these days by participation in a movement, with various degrees of possible participation. This difference in what movement "membership" entails, as well as the different kinds of messages and media available to modern citizens, requires some different nuances in how we define propaganda.

---

[20]Howard Rheingold, *Net Smart: How to Thrive Online*, Cambridge, Mass.: MIT Press (2014), p. 39.

[21]Jacques Ellul, *Propaganda: The Formation of Men's Attitudes*, New York, Vintage Books (1965), p. 61.

With that in mind, I define propaganda as the use of one or more media to communicate a message, with the aim of changing someone's mind or actions *via psychological manipulation*, rather than reasoned discourse. Non-propaganda is not the absence of bias—we're all biased. Propaganda is the (usually purposeful) attempt to hide the bias, to present non-facts as facts, to present facts incompletely or stripped of their essential context, to steer the mind away from the processes of reason that allow us to read through bias critically and to discern facts from fiction, truth from lies. Propaganda can involve *disinformation* (from the Russian *dezinformatsiya*)—a purposeful attempt to deceive or manipulate—or *misinformation*, an inadvertent spreading of falsehoods and fallacies. Online, we often see both working in concert—a purposeful attempt to deceive, shared in such a way that the deceived help to propagate it, but in earnest. I call this multistage propaganda *information laundering*, since the original disinformation is laundered through well-meaning people, whose activity both spreads the message and obscures the source.

With these definitions, we can see how media addiction, invasive data collection, and constant "attentional blink" all prime us to be victims of information operations. The more we spend time on platforms that promote superficial thinking, the less critically we examine the information we engage and the sources from which it comes. The more "social" our media consumption behaviors, the more we let our guard down. The more we shift cognitive gears, the less capable we are of going deep when we need to. And the more data is collected about us, the more sophisticated and personally targeted those operations can be. Harold D. Lasswell writes that propaganda seeks to subordinate others "while reducing the material cost to power,"[22] and our modern, algorithmically based media platforms provide perhaps the greatest opportunity in human history to accomplish that cost reduction.

But all hope is not lost. Ellul writes: "Propaganda renders the true exercise of [democracy] almost impossible" (p. xvi). And as social media feeds the propaganda machine, many critics are sounding the death knell of modern democracy. However, the very tools that facilitate the information operations increasing social polarization and potentially swinging elections *are the same tools that can help us resist.* Ellul also writes that because propaganda dehumanizes and reduces our personal and collective agency, "Propaganda ceases where simple dialogue begins" (p. 6). And where better to start that dialogue than on social media?

I understand. It can be easy to dismiss social media, even the web, and pine for the older forms of human connection (none of which have actually disappeared, by the way). But as new media scholar Clay Shirky puts it,

[22]Cited in Jacques Ellul, *Propaganda: The Formation of Men's Attitudes*, p. x.

"The change we are in the middle of isn't minor and it isn't optional, but nor are its contours set in stone." He continues:

> Our older habits of consumption weren't virtuous, they were just a side-effect of living in an environment of impoverished access. Nostalgia for the accidental scarcity we've just emerged from is just a sideshow; the main event is trying to shape the greatest expansion of expressive capability the world has ever known.[23]

The media landscape is a very different place than it was even just ten years ago. As the constant updates to our social media feeds remind us regularly, nothing is set in stone in this Brave New World. Not yet. We have the opportunity to shape it, as both consumers and producers of the content that keeps those platforms afloat.

But to do that work of resistance, of redrawing the blueprints of media and society, first we need to know how it works. In the next two chapters, we'll dive deeply into how *we* work when we engage information, and then into how the *systems* work. That knowledge, plus the examples—both hopeful and tragic—explored in the latter half of this book, will give us the foothold we need as we seek to solve the propaganda problem.

But for now, I'll leave you with some of the best new media advice I've ever heard.

> Throw some sand into the machinery that automatizes your attention.
>
> —Howard Rheingold[24]

# Summary

In this chapter we've learned that Western capitalism has moved from a commodity-based economy to an attention-based economy. The supply-and-demand equation that increasingly governs the way we interact with information deals with the limited supply and increasing demand of human attention, rather than information, goods, or services.

Algorithmic recommendation engines and social media feeds have been created to help users find the most relevant content and to help media

---

[23]Clay Shirky, "Why Abundance Is Good: A Reply to Nick Carr," *Encyclopaedia Britannica Blog*, July 17, 2008, `http://blogs.britannica.com/2008/07/why-abundance-is-good-a-reply-to-nick-carr/`.
[24]*Net Smart*, p. 50.

producers find the most appropriate audiences. But the ways in which media producers compete for our attention, the amount of personal data mined to make the algorithms work, and the natural way our cognitive systems function all combine to make the modern media landscape ripe for propaganda.

However, if we understand the economy, biology, and technology, we can begin to counter the negative effects and even use the same tools to undo the propaganda problem.

CHAPTER 2

# Cog in the System

## How the Limits of Our Brains Leave Us Vulnerable to Cognitive Hacking

In an attention economy, understanding cognitive psychology gives an informer (or a disinformer) a major advantage in influencing opinion. What attracts attention? How do you hold attention? And how, over time, do you manipulate attention in ways that serve your purposes? These are key questions that we need to find answers to before we can fully grasp the role technology plays in influencing opinion.

## Clickbait: You Won't Believe What Happens Next!

Don't you hate clickbait? Those online titles like "This Father Found a Rattlesnake in his Infant's Crib, and You Won't Believe What Happened Next!" or "27 New Dinner Plans that You Need to Try Before You Die!" As cheap and as cheesy as they sound, these titles are highly engineered, based on research into the clicking habits of millions of users online. Search something like "how to write a viral headline," and you'll find no limit of marketing how-to's telling you to use odd numbers and phrases that create a sense of urgency.


K. Shaffer, *Data versus Democracy*,
https://doi.org/10.1007/978-1-4842-4540-8_2

Why have these titles (and, for that matter, the articles attached to them) become so ubiquitous? Clickbait represents, perhaps better than anything else, life on the modern web. The abundance of free content and the limits of the attention economy have media outlets competing for clicks and the advertising dollars they represent. But because even high-quality information can be readily found with little effort, or money, many media outlets no longer compete to have the best content, but to have the most attention-grabbing content.

That's because in an attention economy, those who thrive aren't those who control information—the ones with access to the biggest libraries or the best content. In an attention economy, those who thrive are those who control attention—those who can master their own attention and those who can attract, and hold, the attention of others.

But just what is attention? And why is it that the one, two, or three things we're aware of at any given moment can so readily dominate the massive store of memories we have tucked away in our brains?

## Mapping the Cognitive System

Think of your computer. It stores data in several different places—the hard drive, RAM, the processor's cache—each with its own function, and with its characteristic strengths and weaknesses. The hard drive is the largest data store. Even my 13-inch laptop can hold a terabyte of data on its hard drive. But compared to other data stores, it is also the slowest and least efficient when it comes to accessing, writing, and erasing data.

RAM (random access memory), on the other hand, is considerably smaller, but also faster and more flexible. Because of its power, and its limited size, applications are constantly vying for access to it, and memory management is a key element of optimizing the performance of any app or algorithm.

Even smaller—and more powerful—is the cache memory that lives on the processor itself. This is where the action happens. But it is severely limited according to what the processor itself is capable of. That's because data storage means nothing if you can't *do* something with it. The computer's data processing capability ultimately sets the limits for what we can do with data.

This hard drive, RAM, cache, processor model is a helpful analog for the human cognitive system. It's an oversimplification, of course, but it's a helpful starting point, especially if we want to know how the brain interacts with (big) data.

In many ways, the brain is one big hard drive. There's a reason that we refer to a computer's data storage as "memory," after all. Both the brain and the hard drive store information in small, hierarchically organized units, which are activated, transferred, and copied via electrical signals. Some cognitive scientists even use the language of "bits" to describe the information our brains process.[1]

But rather than having separate organs for large/slow storage, fast/efficient storage, and data processing, the brain does it all in one massively complex, and flexible, organ. However, the brain's capabilities *are* separated analogously to how a computer works. (Well, technically, it's the *computer* that works like the *brain*.) The largest share of the brain's storage capability is less efficient and not directly accessible to the "processor." This is typically called *long-term memory* (LTM)—where we keep memories of life events (*episodic memory*), physical skills and processes (*procedural memory*), and important information like the identity of friends or the meaning of words (*semantic memory*).[2] It's massive. Each human brain contains the informational complexity of the entire universe of stars, planets, and galaxies.[3] But we can't access it all at once. That's just too much for the brain to handle. We need something to help us manage all of that data.

## The Limits of Conscious Attention

That's where working memory comes in. Working memory is the combination of what scientists call the "central executive" (the brain's CPU) and several independent resources for short-term, high-efficiency storage, often called *short-term memory* (STM—the brain's RAM).[4]

Think of it this way. Every memory ever formed—every event you've experienced, every word you know, every friend and family member, every skill you've formed—is stored somewhere in your long-term memory (LTM). Because it takes energy (in the form of electrical signals) to access these memories, only some of those memories are "activated," like an application launched and ready to use, or books taken off the shelf and ready to be read.

[1]This is based on the "Shannon–Weaver equation," developed by Claude Shannon and Warren Weaver in *The Mathematical Theory of Communication*, Urbana, Ill.: University of Illinois Press (1949).

[2]Alan D. Baddeley, *Human Memory: Theory and Practice*, East Sussex: Psychology Press (1997), p. 29ff.

[3]Christof Koch and Patricia Kuhl, "Decoding 'the Most Complex Object in the Universe'," interview by Ira Flatow, *Talk of the Nation*, NPR, June 14, 2013, `www.npr.org/2013/06/14/191614360/decoding-the-most-complex-object-in-the-universe`.

[4]Alan D. Baddeley, *Human Memory: Theory and Practice*, East Sussex: Psychology Press (1997), p. 29ff.

This is short-term memory (STM). When something is activated and placed in STM, it's only there for a short amount of time. That is, unless we do extra work to keep it alive. (Cognitive musicologists call that work "rehearsal," but it's actually a lot like renewing a library book before it's due.) When memories are in STM, we can do more with them—put them in order, forge relationships between them, and build higher-level groupings of them (called *schemas*). But only some of what we've called into STM forms conscious awareness—the small amount of information being processed *right now*. That's the information that makes up the focus of our *attention*.

## The Triggers of Attention

Attention is incredibly expensive. Such a high level of neural activation, not to mention the processing that goes on, requires large amounts of energy. That energy comes at a premium. And so the brain keeps all but a tiny window of time and data at a lower state of activation, away from the costly power of consciousness.

Of course, such an efficient and orderly system requires structure—rules that decide what we pay attention to, and when. When it comes to external stimuli—sights, sounds, smells—there are many rules of prioritization based on millennia of natural selection.

As we've already discussed, our genetic code was largely written by our ancient ancestors as they fought for survival.[5] Thus, things that meant life or death on the savannah 150,000 years ago are more likely to command our attention today. If someone sneaks up behind you and makes a loud, sudden noise, you'll jump, suck in extra oxygen (gasp), your hair will stand on end (the vestige of our ancestors making themselves look bigger and fiercer than they are), and your heart will speed up (preparing to deliver that extra oxygen to your muscles, whether they need it for fight or for flight). Perhaps surprisingly, if they sneak up on you again, but this time they tell you first, some of those reflexes will still kick in, *even though you knew it was coming*. Some of our ancestors survived because these reflexes were so fast and reliable that the brain could never turn them off. (In fact, some of these reflexes, called "spinal reflexes,"[6] are directed by the spinal cord, because the speed-of-light trip of our nerve impulses all the way to the brain and back simply takes too long!) And those ancestors who survived passed their saber-tooth-tiger-evading genes onto us.

[5]Jerome H. Barkow, Leda Cosmides, and John Tooby (eds), *The Adapted Mind: Evolutionary Psychology and the Generation of Culture*, New York, NY: Oxford University Press (1992).
[6]James Knierim, "Spinal Reflexes and Descending Motor Pathways," in *Neuroscience Online*, University of Texas McGovern Medical School, accessed February 8, 2019, `https://nba.uth.tmc.edu/neuroscience/s3/chapter02.html`.

Other memories—those less critical to our ancestors' survival—can be recalled on purpose. Who is that? What's their phone number? I came into this room for something, what was it?

Still others feel more natural, more automatic. We may have to actively recall our new colleague's name on their second day of work, but it doesn't take any work at all to remember our spouse's, partner's, or child's name. That's because inside our "big data" brain lies the world's most advanced predictive analytics engine. Data we access often remains more "activated" than data seldom accessed, like the spices that always end up in the front and center of the cabinet, despite our attempts to alphabetize. Memories that are related to other memories we've already brought to a higher state of activation will likewise raise in activation. This is why we might not remember that password or that phone number out of the blue, but when we sit at the keyboard or pick up the phone, it just comes to us. Cognitive scientists call this *priming*.

Priming is an essential part of how our brain manages our memories. It anticipates our needs, delivers what we need in time to respond quickly (just in case), and it does so while keeping energy usage low. But because priming is such a key element of how we manage our own attention, it can also be used by others, in combination with hard-wired evolutionary rules for processing stimuli, to command our attention and to control, at least in part, the way we think about the world.

## Familiarity Breeds Believability: The Role of Unconscious Memory

We are constantly evaluating everything we perceive. Is it safe or dangerous? Is it good or bad? Is it exciting or boring? Did we expect it, or was it a surprise? These evaluations feed into what we experience as emotions. When something is surprising and dangerous, the result is fear. If that surprising and dangerous perception holds our attention for a long time, the result is terror. If, on the other hand, that surprising thing turns out to be not very dangerous at all, the result might be laughter—or as cognitive scientist David Huron calls it, "pleasurable panting" (what you do with all that oxygen your immediate fear responses sucked in).[7] If something is predictable and harmless, the result might be boredom, especially if that something keeps going on for a long time (think of a concert or a lecture you found uninteresting, but had no way to escape). On the other hand, if something is difficult to predict, but also harmless, the result can be disorientation, and if it goes on for a long time,

[7]David Huron, *Sweet Anticipation: Music and the Psychology of Expectation*, Cambridge, Mass.: The MIT Press (2006), p. 26.

frustration.[8] This is the core of the creepy-crawly feeling some people get when listening to avant-garde music, or the repulsion they feel when they experience other avant-garde art forms. It's just music, or a painting, or a sculpture, but the fact that it arrests our attention so strongly, for so long, but the brain doesn't know how to make sense of it, that's a very bad thing, evolutionarily speaking. And it makes perfect sense that our ancestors would have evolved an emotional response to such situations that would cause them to try and avoid them in the future.

Emotion can be a very complex thing. Entire books have been written about the cognitive science of emotion. But for our purposes in understanding how humans interact with online media, we'll cover just a few key elements. Things that affect our ability (or inability) to recognize disinformation, misinformation, and "fake news" online, as well as things that promote social polarization. We'll start with something simple, yet powerful: the *mere exposure effect*.

Because of our evolutionary past, we are always assessing whether something we encounter or perceive—a stimulus—is positive or negative. (Scientists call the positivity or negativity of a stimulus its *valence*.) Our life-or-death evolutionary history means that we make some of these assessments very quickly. These snap judgments can save our lives, but if they happen unconsciously in certain social settings, they can form the stuff of overgeneralizations, stereotypes, and even lead to sexism and racism.

One factor that contributes to a positive affect when we encounter a stimulus is the ease with which our brain can process what it perceives—called *perceptual fluency*. If the brain can process, understand, and evaluate the stimulus quickly and without difficulty, the result is a positive affect or emotion. This is why, for example, notes right next to each other on the piano (and the cochlea) are "dissonant" rather than forming a pleasant chord.[9] Things that make it harder for the brain to process—similar colors, sounds, or appearances—make it hard to distinguish perceived items and place them into categories, the way the brain likes.

Differences in color, musical pitch, and the like are, for the most part, hard-wired differences. We may be able through artistic and musical training to make finer judgments than the average person, but not much finer. There are physical limitations in our inner ear and on our retina that determine how finely we can develop our perceptual abilities.

But other differences are learned. We learn language, bodily gestures, faces, and skills through repeated exposure. Because the brain conserves energy by

---

[8]Patrick Colm Hogan, *Cognitive Science, Literature, and the Arts*, New York: Routledge (2003), 9–11.
[9]R. Plomp and J. M. Levelt, "Tonal Consonance and Critical Bandwidth," in *Journal of the Acoustical Society of America* 37 (1965), pp. 548–60.

preactivating parts of memory that it anticipates us needing, and because it's more likely to anticipate something that we encounter regularly than something we encounter rarely (let alone something entirely new), the brain preactivates the resources needed to parse *familiar* things far more readily than the resources needed to parse *unfamiliar* things. As a result, the brain processes familiar things faster and more easily than it processes unfamiliar things. Finally, since fast, easy processing is associated with a positive reaction, familiar things tend to lead to more positive emotions than unfamiliar things, all else being equal.

This makes sense. When a piece of music ends the way we might expect, that feels good. No surprises. Everything is in its right place. And that's true even when we don't know enough music theory to explain what "should" happen or why. It's why a clean house or a well-kept garden makes us feel relaxed, even at home—when everything is "as it should be," our brain has an easy time making sense of the environment, even if it's a new one. (Remember, aeons of evolutionary history mean we're always looking for danger! A clean environment, with everything in its place, makes it easier for us to get a handle on what's out there and makes it safe to let our guard down.)

A number of scientific studies have demonstrated these principles at work. In one particularly famous study, researchers showed their subjects (who had no knowledge of the Chinese language) a series of Chinese characters and asked them to provide an adjective that they thought the character might stand for. Characters that the subjects had already seen earlier in the study tended to be attributed adjectives with a more positive connotation than characters the subjects were seeing for the first time.[10]

While it seems like a simple enough study, it's actually rather remarkable that previous exposure to a character can lead to a discernible and statistically significant increase in the positivity associated with that character. And this kind of phenomenon happens regardless of whether the subject recognizes the object! That's because we build up this "statistical" awareness unconsciously.

I've always been taken by music cognition studies, and they provide some of the most interesting insights into these kinds of phenomena. A number of studies have been done to measure how humans internalize the patterns in the music they hear.[11] From testing infants' musical expectations by tracking the movement of their head or eyes, to asking people to sing the note they think might come next, to having subjects rate how well a note "fits" with a

---

[10]Jennifer L. Monahan, Sheila T. Murphy, and R. B. Zajonc, "Subliminal Mere Exposure: Specific, General, and Diffuse Effects," *Psychological Science* 11/6 (2000), 462–66.
[11]See, for example, Aniruddh D. Patel, *Music, Language, and the Brain*, Oxford: Oxford University Press (2008); Huron, *Sweet Anticipation*; Patrick N. Juslin and John A. Sloboda, *Music and Emotion: Theory and Research*, Oxford: Oxford University Press (2001); and the journal, *Music Perception*—just to name a few.

melody, to giving subjects money and asking them to bet on which note will come next, these studies come at the question from just about every angle. But they generally come to the same conclusion: humans are really good at learning patterns through repeated exposure and making predictions and judgments in accordance with the patterns we learned, even if we can't explain the reason behind those judgments.

One study[12] presented Western musicians with a melody one note at a time, but the melody was from a style that was unfamiliar to all of them—Balinese *gamelan* music. (If you've never heard, or *seen*, Balinese gamelan music performed before, find a few videos on YouTube. It's incredible.) This music not only follows different rules than Western classical music, pop music, jazz, blues, country, bluegrass, you name it. It is based on an entirely different system of scales than Western music. You cannot play Beethoven on a gamelan, nor could you perform a gamelan piece on a piano. It simply wouldn't work. That, plus the musicians' lack of exposure to the style, made it an excellent means to test how a human, in this case an expert Western musician, might approach "learning" a completely new musical style from scratch.

The experimenters played first one note of the gamelan melody and asked each subject to predict what note they thought most likely to come next. The guesses were just that: guesses. No better than chance at predicting the next note. After two notes, three notes, still fairly wild guesses. But as the melody progressed, and the subjects had more context for their predictions, their predictions improved. To be clear, they were not learning the *rules* of the musical style. In fact, they missed some core elements of the musical style completely. But they were learning the *statistics* of the style, the basic tendencies and proportions. After hearing enough of the melody, if the melody so far had more *C's* than *E-flats*, they would predict C more often than E-flat. And by the end of the melody, though they weren't exactly gamelan experts, their predictive powers were noticeably better than chance.

This ability to learn patterns quickly, with detail, and unconsciously is a core part of our humanity. It's a large part of how we learned our first language. We weren't born with an English gene, or a Japanese gene, or a Farsi gene. We didn't sit down at the blackboard for lessons on how to move the lips, jaw, and tongue to say "mama" or "dada." We are born with a genetic predisposition to learn, and our brain hones in on the patterns we hear most often and does all it can to emulate them—and to improve upon our ability to emulate them. The ability to make predictions and evaluations based on those patterns is also part and parcel to being human—even leading to our deepest flaws, including responding negatively, even violently, to the unfamiliar. Our racism, sexism, homophobia, transphobia, xenophobia, ... they all stem from a

[12]Huron, *Sweet Anticipation*, pp. 53–55.

negative reaction to the unfamiliar, and our ancient association of the unfamiliar with danger and threat.

Now, of course there are other factors in how we appraise things. As discussed previously, the associations our brains have formed over time, as we have repeatedly engaged things *in context* (rather than in an experiment), also play a huge role. But all things being equal, familiarity, even unconscious familiarity, leads to a more positive appraisal of an object, a person, or an idea.

This core psychological principle forms the basis of much practice in advertising. Think of product placement. The subtle inclusion of a product in a film or television show, often paid for just like a commercial, by making that product slightly more familiar, slightly more preactivated, primes you to appraise it more positively when deciding whether or not to purchase it in the future. They might hit you over the head with it—like Reese's Pieces in the movie E.T.—or it might be more subtle—like the label on the can of soda sitting on the edge of the table just inside the frame—but the principle remains the same: repeated engagement leads to positive appraisal, all other factors being equal.

One wildly successful example from the world of advertising is the white Apple earbud.[13] Part of the marketing around the iPod itself was its small size, even its hideability ("1,000 songs in your pocket"). The iPod couldn't fulfill its own identity as a stealthy device while also being its own marketing campaign—"Look at that! I must have one!" However, at that point in time, white earbuds were extremely rare. By bequeathing a pair of standout white earbuds with every new iPod (and later, iPhone), the device could keep its low profile, tucked away in every listener's pocket, while the earbuds could play the marketing game.

Apple capitalized on this with its visual advertising campaign. Remember the commercials, posters, and magazine ads with a black silhouette of a person dancing in front of a background made up of a single, solid, bright color?[14] The person dancing was always holding an iPod in their hand (also part of the black silhouette). The only other element in the photo or video was the pair of Apple earbuds. White. Stark against the black and fuchsia/orange/green. The visual focal point of the advertisement. The earbuds highlighted in this ad campaign and the earbuds seen on the street or in the train reinforced each other, generating a sense of ubiquity and raising the familiarity of the symbol. Never before had the color of the wire between someone's ear and their

---

[13]Sasha Geffen, "The iPod May Be Dead, but Those Iconic Ads Still Shape the Way We See Music," *MTV News*, published Mary 12, 2016, `www.mtv.com/news/2879585/ipod-ads-in-music-culture/`.

[14]"2003: Apple Releases its Silhouette Campaign for iPod," *The Drum*, March 31, 2016, `www.thedrum.com/news/2016/03/31/2003-apple-releases-its-silhouette-campaign-ipod`.

pocket primed us to think (usually positively) about a specific brand. But by picking something rare yet visible and reinforcing it with their visual advertising materials in a way that took advantage of the way our brain works, Apple was able to do just that.

Now, familiarity itself is no guarantee of a positive response. Those same white earbuds, once they reach a point of market saturation, can start to stand out too much. Some people recognize the gimmick. Others simply want to avoid being like everyone else. And so *how much is too much?* is a question advertisers have been asking themselves for decades. A single Superbowl ad isn't worth the expensive investment if you have no marketing budget left afterward. But the same commercial seen during every commercial break of every show is also counterproductive. Remember that when something is too familiar, but arrests our attention anyway, the waste of resources can lead to a negative reaction. Finding the sweet spot (which is different for every individual) is a major part of the advertising game.[15]

Also remember that when two things are associated in our memory, activating one will activate the other at the same time.[16] So while repeated exposure to white earbuds might boost familiarity with both the earbuds and the device they come with, that only leads to a positive appraisal if you don't already have a negative association with the earbuds, the iPod, or the Apple corporation. This is why after major scandals, companies often rebrand, either renaming a product or renaming the company altogether.

I imagine you can already see potential implications here for social media and big data analytics. If seeing something too often is bad, not often enough is ineffective, and the sweet spot is different for each person, then individualizing how often someone sees an ad would optimize its effect. Likewise, if associations in a person's memory affect the positive or negative appraisal they have when seeing or hearing something, and if different people have different associations, then individualizing the content of an ad would likewise optimize its effect.

Of course, individuals aren't *entirely* unique. Though you and I might react differently to the same ad, we might react similarly *enough* to someone else that showing us the same content, on the same schedule, would have more or less the same effect. And so advertisers have been grouping us together for ages—single men ages 18–25, tweenage girls from upper-middle-class families, retired working-class widowers on medicare, city dwellers, suburbanites, homesteaders, you name it. Marketers divide people into groups based on demographics, do their market research, and target ads to groups accordingly.

---

[15]Erin Richards, "Cognitive Efficiency Determines How Advertising Affects Your Brain," in *Science 2.0*, December 9, 2008, `www.science20.com/erin039s_spin/cognitive_efficiency_determines_how_advertising_affects_your_brain`.
[16]Donald O. Hebb, *The Organization of Behavior*, New York: Wiley & Sons (1949).

This is why there are a lot more toy commercials during children's television shows than during daytime soap operas, for example.

But TV ads can't really target as precisely as advertisers might like. Are the people most likely to buy product X or vote for candidate Y really all watching the same show at the same time in the same broadcast/cable/satellite market? Surely our identities are not defined and clustered according to our television viewing habits, our morning-commute radio preferences, and our magazine subscriptions!

The internet drastically altered this equation. Web technology allows advertisers (and purveyors of all kinds of information) to microtarget very small groups of people—even individuals, if they *really* know what they're doing. (Targeting individuals with ads is forbidden on most major online platforms, but there are ways around that for the more talented, if less ethical, information operatives.) If the optimal content and timing is *individual*, if the groups of people with similar preferences and associations do not line up with TV show viewership or commute schedule, then targeting online starts to make more sense. And in a media environment where every web site visited, every link clicked, every image or video viewed, every song listened to, every like, every retweet, every :rage: emoticon is logged *and available to advertisers* (at least as filters), well ... then we can really start to put our knowledge of human cognition to work in our marketing practices.

In a world where every "sponsored post" is effectively an advertisement, advertising is not simply about selling things. It's about political campaigns. It's about journalists and freelance writers promoting their work. It's about indie musicians finding an audience. It's about inventors crowdfunding their prototypes. Anyone who wants to communicate information or effect human behavior, and who has the money to support it, can target individuals in finely grained ways without ever scraping a single bit of data themselves. And in spite of all the limitations put in place by the platforms, unethical operators can do some really shady things. (More on *that* in the chapters to come.)

# Summary

Let's sum up what we've learned. The brain is a large and complex information system, with optimized storage solutions and a powerful predictive analytics engine. It's the real intelligence that the best AI systems are still trying to catch up to. The result of the brain's optimized information storage and predictive analytics algorithms is a bottleneck called attention—the small amount of information we can "think about" at any given time, including memories and signals from our five senses. Biologically, attention is incredibly powerful, but also incredibly expensive, and therefore incredibly limited. Over the course of human evolution, we have evolved processes for prioritizing the information that makes it into conscious attention, including both hard-wired rules by which certain kinds of stimuli will always take priority as well as the capability to *learn* new priorities and connections that govern our attention.

One of the main things we learn is statistical frequency (familiarity). Memories we engage frequently stay "preactivated," ready for easy access. We also learn relationships between memories and stimuli, and when we interact with one thing in our memory, all of the things our brain has related it to will also raise in their level of activation, making it easier for us to process what we predict.

We also know that, all else being equal, when a perception is easy to process, we form a positive emotional response. And that unconscious familiarity can be a powerful driver of positive responses to things we perceive.

Of course, the degree to which something is familiar, whether or not we are aware of that familiarity, and what associations (positive and negative) we might have already made with that thing is highly individual. And while advertisers have been dividing us into groups for decades, and serving up ads likely to be the most effective for the particular group(s) we belong to, the optimal solution will be individualized.

But there are a few problems with completely individualized advertising. First, in order to work, it requires access to tons of data about us as individuals—much of which we likely view as private. This includes knowledge of what we've seen, heard, smelled, tasted, touched, in what contexts we've interacted with it, and how we reacted to it in those different contexts. Where relevant data is absent, it requires a means of filling in the gaps (more on that in Chapter 3). Assuming advertisers have the data needed to manipulate us in the way they desire, this completely individualized advertising can have effects on both our behavior and our psychology that were *not* intended, or even thought of, by the advertisers. And the effects—both intended and unintended—are more difficult to regulate, as the groups of people who see the same advertisements are far smaller and more diffuse than TV markets or subscribers to a particular magazine. Finally, what happens when the (private) data collected to make this hyper-targeted advertising effective is sold, leaked, or hacked?

We'll unpack these questions in Chapter 4 and beyond. But first, let's get our heads around how the attention economy, our cognitive limitations, and personalized, targeted content work together in the increasingly central feature of modern human society: the algorithmic news feed.

CHAPTER

3

# Swimming Upstream

## How Content Recommendation Engines Impact Information and Manipulate Our Attention

As we have shifted from an information economy to an attention economy in the past two decades, we have almost simultaneously shifted from mass media to social media. The drastic increase in available media necessitates a way for individuals to sift through the media that is literally at their fingertips. Content recommendation systems have emerged as the technological solution to this social/informational problem. Understanding how recommendation system algorithms work, and how they reinforce (and even exaggerate) unconscious human bias, is essential to understanding the way data influences opinion.

## What's New?

Online advertising can be uncanny.

My wife was on Facebook a while back and noticed an odd advertisement. Facebook was suggesting that she might be interested in a set of patio seat cushions. *That's funny*, she thought, *I was just on the phone with my mom, and she told me she was out looking for just such a cushion!* A couple weeks later, the


K. Shaffer, *Data versus Democracy*,
https://doi.org/10.1007/978-1-4842-4540-8_3

same thing happened with a table lamp from a store she hadn't shopped at in years, but her parents had just been to. I asked her if she checked Facebook from her parents' computer when we visited the previous month, and, of course, the answer was yes. She was sure to log out when she was done, but the cookie Facebook installed into her parents' web browser was still there, phoning home, while her parents shopped online for seat cushions and lamps. A few web searches (and credit card purchases) later, her parents' shopping had been synched to her profile on Facebook's servers. So as her parents perused household items, she saw ads for similar items in her feed.

In that case, Facebook made a mistake, conflating her online identity with that of her parents. But sometimes these platforms get it a little *too* right.

You may have heard the story in the news a few years ago. A father was livid to receive advertisements from Target in the mail, addressed to his teenage daughter, encouraging her to buy diapers, cribs, baby clothes, and other pregnancy-related items. "Are you trying to encourage her to get pregnant?" he reportedly asked the store manager. A few days later, the manager phoned the father to reiterate his apology. But the father told him, "It turns out there's been some activities in my house I haven't been completely aware of. [My daughter]'s due in August. I owe you an apology."[1]

It turns out that Target, like many retailers today, was using customers' purchasing history to predict future purchases and send them advertisements and individualized coupons based on those predictions, all in an effort to convince existing customers to turn Target into their one-stop shop.

But it's not just advertisers who feed user data into predictive models that generate hyper-targeted results. In her book *Algorithms of Oppression: How Search Engines Reinforce Racism*, Safiya Umoja Noble tells the story of a harrowing Google search. Looking for resources on growing up as a young woman of color in rural America to share with her daughter and her friends, Noble picked up her laptop and searched for "black girls" on Google. She writes:

> I had almost inadvertently exposed them to one of the most graphic and overt illustrations of what the advertisers already thought of them: Black girls were still the fodder of porn sites, dehumanizing them as commodities, as products and as objects of sexual gratification. I closed the laptop and redirected our attention to fun things we might do, such as see a movie down the street.[2]

---

[1]Charles Duhigg, "How Companies Learn Your Secrets," *The New York Times Magazine*, published February 16, 2012, `www.nytimes.com/2012/02/19/magazine/shopping-habits.html`.

[2]Safiya Umoja Noble, *Algorithms of Oppression: How Search Engines Reinforce Racism* (New York University Press, 2018), p. 18.

Noble's book is full of other example searches that, like the ones we examined in Chapter 1, return predictive results based on the biases and prejudices of society—image searches for "gorillas" that return pictures of African Americans, searches for "girls" that return older and more sexualized results than searches for "boys," etc. In these cases, the results are not "personalized"—Noble was clearly not looking for the pornography that Google's search engine provided her in the first page of search results. The results take the biases of the *programmers* who created the algorithm and the biases of the *users* who search for "black girls" and, apparently, click on links to pornography more often than links to web sites containing resources meant to empower young women of color. These biases then return results predictive of what most users "want" to see.

Google has since addressed problems like this. Now, a search for something neutral like "black girls" on my computer (with SafeSearch off) returns sites like "Black Girls Rock!" and "Black Girls Code," as well as images of black women fully dressed. (Note that only a minority of the images included actual *girls*, as opposed to grown women. Progress is progress, but we still have work to do...)

But what happens when the search itself isn't neutral? What if the search is explicitly racist?

Before terrorist Dylann Roof committed mass murder in Charleston, South Carolina, in 2015, he left behind a "manifesto." In that manifesto he *claims* (and let me be clear, I take the claims made in any terrorist's manifesto with a humungous grain of salt) that he was motivated to pursue his "race war" by a Google search. According to Roof, he searched for "black on white crime," and the resulting pages provided him with all the information he needed to be self-radicalized.[3] Google has since made updates that correct searches like this, as well, to the point that three of the top ten results for that search on my computer today lead me to official U.S. government crime statistics. But in 2015, people looking for government crime statistics broken down by demographic rarely searched using terms like "black on white crime" or "black on black crime." Those were phrases most often used by racists, both in their searches and on their web sites. That very search reflects the likely racism of the search engine user, and even a "neutral" algorithm would likely predict racist web sites to be the most desirable results.

Now, let me be clear. When someone who clearly has the capacity to kill nine perfect strangers in a church at a Bible study searches for "black on white crime," that person is already well on the route to radicalization. Dylann Roof was not radicalized *by* a Google search. But given all search engines' ability to return biased results for a "neutral" search like "black girls," it's important to

[3]"Google and the Miseducation of Dylann Roof," Southern Poverty Law Center, published January 18, 2017, www.splcenter.org/20170118/google-and-miseducation-dylann-roof.

note how biased the predictive results can be when the input itself is biased, even hateful.

The bias-multiplication effect of search engine algorithms is perhaps seen most starkly in what happened to *The Guardian* reporter, Carole Cadwalladr, a few years ago. She writes:

> One week ago, I typed "did the hol" into a Google search box and clicked on its autocomplete suggestion, "Did the Holocaust happen?" And there, at the top of the list, was a link to Stormfront, a neo-Nazi white supremacist web site, and an article entitled "Top 10 reasons why the Holocaust didn't happen."[4]

Perhaps even more scandalous than the anti-Semitic search results was Google's initial response. Google issued a statement, saying, "We are saddened to see that hate organizations still exist. The fact that hate sites appear in search results does not mean that Google endorses these views." But they declined at the time to remove the neo-Nazi sites from their search results.

Eventually, Google *did* step in and correct the results for that search,[5] so that today a Google search for "Did the holocaust happen?" returns an entire page of results confirming that it did, in fact, happen and explaining the phenomenon of Holocaust denial. In my search just now, there was even an article in the first page of results from *The Guardian* about the history of this search term scandal.

Google's initial response—that the algorithm was neutral, and that it would be inappropriate to alter the search results manually, regardless of how reprehensible they may be—hits at the core of what we will unpack in this chapter. This idea of *algorithmic neutrality* is a data science fallacy. Developing a machine learning model always means refining the algorithm when the outputs do not match the expectations or demands of the model. That's because mistakes and biases always creep into the system, and sometimes they can only be detected once the algorithm is put to work at a large scale.

In the case of search results, though, the model is constantly evolving, as millions of searches are performed each day, each one altering the dataset the model is "trained" on in real time. Not only will a search engine algorithm propagate the conscious and unconscious bias of its *programmers*, it is also open to being gamed consciously or unconsciously by its *users*.

---

[4]Carole Cadwalladr, "How to bump Holocaust deniers off Google's top spot? Pay Google," *The Observer*, published December 17, 2016, `www.theguardian.com/technology/2016/dec/17/holocaust-deniers-google-search-top-spot`.

[5]Jeff John Roberts, "Google Demotes Holocaust Denial and Hate Sites in Update to Algorithm," *Fortune*, published December 20, 2016, `http://fortune.com/2016/12/20/google-algorithm-update/`.

Left unchecked, a content recommendation algorithm is a bias amplifier. That is as true for social network feeds as for search engines. Understanding how they work on a general level can help us better understand online self-radicalization, partisan polarization, and how actors can manipulate platforms to advance their agendas on a massive scale. It can also help us think more critically about the information we encounter in our own social feeds and search results, and give us the foothold needed to effect positive change in the online information space.

So let's dive in...

## How the Stream Works

Social media and search algorithms are important trade secrets. Outside of the developers who work on those algorithms directly, no one can know with 100% certainty how they work in all the gory details. However, equipped with a knowledge of how machine learning algorithms work in general, alongside experiences of seeing the inputs and outputs of the models, we can reverse engineer a fair amount of the details. Add that to research papers published by the data scientists behind these models, and we can paint a fairly good picture of how they work and the effects they have on society.

Let's use a search engine as an example. When someone performs an image search, the search algorithm, or model, considers a number of inputs alongside the search term in order to determine the output—the result images in an order that surfaces the images most likely to be clicked at the top of the list.

In addition to the search query itself, the model considers information from my user profile and past activity.[6] For example, if my profile tells it that I currently live in the United States, I'll likely see different results than if my profile tells the model I live in India or Venezuela. My profile for most platforms also contains information about my gender, age, education, marital status, race, etc., either because I provided that information or because the platform inferred it from my activity.[7] Inferred data is, of course, imperfect—one of my Google accounts thinks I'm male, another thinks I'm female, and both have wildly misidentified my taste in music and sports—but it is used to personalize content alongside user-provided data.

Depending on the platform, various kinds of past activity are weighed to determine the content recommended by the algorithm. For a search engine, that may be past searches and the results we clicked. For advertising platforms,

[6]"How search algorithms work," Google, www.google.com/search/howsearchworks/algorithms/.

[7]You can find some of the information Google has inferred about you from your online activity at https://adssettings.google.com/.

it may include purchase made at partner stores, products viewed on partner e-commerce web sites, or retailers in our area frequented by many of our social media contacts. For a social network content feed, it likely includes the kinds of content we post and the kinds of content we "engage" most, measured by likes/favorites, comments, clicks, and in some cases, the amount of time we spend looking at it before we scroll by.[8]

But there's a third, and probably the most important, category of inputs that help the model determine the content we see: *data that has nothing to do with us.*

This, of course, includes information about the content under consideration—the data and metadata that makes that content unique, general information about its popularity, etc. But it also contains information about the activity of *other users.*[9]

Data from other users is key for any content recommendation engine. Without that data, the recommendation would only have past user activity as the basis for determining the best content to surface next. Because every person and situation is unique, that small amount of data would provide very little context in which to make a decision. In many cases, it would be flying blind, with no relevant data to draw on to make a prediction. As a result, the algorithm's recommendations would be rather poor.

Think about it this way. In many respects, a content recommendation engine is like a dating app, except instead of matching a person with another person, it matches a person with content. Only in the case of a dating app, it's possible to require a minimum amount of data from a user before proceeding, ensuring that user matches can be filtered and ranked according to complete profiles.[10] The same is true for the algorithms that assign medical school graduates with residencies.[11] But in the case of search engines, social networks, and even "personalized" educational apps, the algorithm needs to filter and rank content that's unlike any the user has engaged with before. The solution is to join the user's data with that of other users in a process called *collaborative filtering.*

To understand collaborative filtering, let's consider a simplistic profile for musical taste. Suppose there were a number of musical features that were

---

[8]I don't want to call out any single developer or company here, but a web search for "track user scrolls on a page" returns a number of solutions for web developers who want to track what web site visitors scroll by and what they don't.

[9]This is even true for "personalized" education apps. See "Knewton Adaptive Learning: Building the world's most powerful education recommendation engine," Knewton, accessed July 24, 2017, `https://cdn.tc-library.org/Edlab/Knewton-adaptive-learning-white-paper-1.pdf`.

[10]This type of model is true for survey-based dating apps like eHarmony and OkCupid, not for behavior-based apps like Tinder.

[11]"How the Matching Algorithm Works," The National Resident Match Program, `www.nrmp.org/matching-algorithm/`.

present or absent in any given musical track: distortion guitar, grand piano, operatic soprano vocals, major key, minor key, fast tempo, slow tempo, improvised instrumental solos, etc. Some of these are binary (either there are bagpipes or there aren't), and others exist on a scale (the relative volume of that distortion guitar, the actual tempo measurement in *beats per minute*, etc.). The more features the model contains, the more refined predictions it can make. But the more features the model contains, the more data it needs to make those predictions. And the more likely it is that at least some of that data is missing from a user's profile.

This leads to a paradox: to ensure a good user experience, especially when trying to hook new users, the app needs to collect as much relevant data as possible before choosing the song. On the other hand, to ensure a good user experience, the app needs to serve up good songs with as little delay as possible—no time for onboarding.

In order to provide a quality user experience, the algorithm needs a way to make good predictions without a complete profile. That's where collaborative filtering comes in.

Collaborative filtering provides a way to fill in the gaps of a user's profile by comparing them with other users.[12] The theory behind it is this: if user A and user B have similar tastes for the features they both have data on, they are likely to have similar tastes for the features where one of them is missing data. In other words, if a friend and I both like distortion guitar, fast tempos, and dislike jazz, then my tastes about various classical music features will be used to make music recommendations for that friend, and their taste about country music will be used to inform my music recommendations. Our incomplete but overlapping profiles will "collaborate" to "filter" each other's musical recommendations—hence the name.

With millions (or, in Facebook's case, *billions*) of users contributing data to the same model, the algorithm can theoretically cluster all of those users into tens or hundreds of thousands of "collaborative" groups, whose profiles will be combined into one super-profile. That super-profile can be used to filter and rank potential content for all of those hundreds or thousands of users in the group, and each one of them will encounter a "personalized" experience—one that is different from anyone else they know.

The clusters can operate at various levels of detail. When I join Pandora and select the preloaded New Wave station as my first listening experience, it serves up songs based on the taste of other users with a New Wave station in their library. But as I give a thumbs up to The Cure, Depeche Mode, and

---

[12]Albert Au Yueng, "Matrix Factorization: A Simple Tutorial and Implementation in Python," quuxlabs, published September 16, 2010, `www.quuxlabs.com/blog/2010/09/matrix-factorization-a-simple-tutorial-and-implementation-in-python/`.

A Flock of Seagulls and a thumbs down to The Smiths and most songs by Duran Duran, it starts to align me with a smaller cluster of listeners who prefer "Space Age Love Song" to "Girls on Film."

To summarize, models make better predictions when they have access to more data—both more unique *observations* and more *features* for each observation. However, the more features a model takes into account, the more likely it is that each user's profile will be missing critical features. So an additional algorithmic model will cluster users together according to their similarity of *known* features, so that a super-profile can be created which will provide data to fill in the *unknown* features. The predictive model then uses these newly complete profiles to generate content recommendations. This leads to a "personalized" experience that, in many ways, amounts to a cluster-based experience. They have a similar experience to many other users, they simply don't interact with those other users, so their experience *feels* unique.

## Bias Amplifier

Think back to the image searches we performed in Chapter 1: *doctor*, *nurse*, *professor*, *teacher*, etc. As discussed in that chapter, the feedback loop between the algorithm and the humans that use it takes already existing human biases and amplifies them. With a bit more understanding of how collaborative filtering works, we can now add some nuance to that feedback loop.

Figure 3-1 illustrates the feedback loop(s) by which human biases are amplified and propagated through unchecked algorithmic content delivery. When a user performs a search, the model takes their search terms and any metadata around the search (location, timing, etc.) as inputs, along with data about the user from their profile and activity history, and other information from the platform's database, like content features and the profiles and preferences of other similar users. Based on this data, the model delivers results—filtered and ranked content, according to predictions made about what the user is most likely to engage with.[13]

[13]Note that I didn't say "most likely to be satisfied with." Attention is the commodity, and engagement the currency, in this new economy. Taste is much harder to quantify, and thus to charge advertisers for.

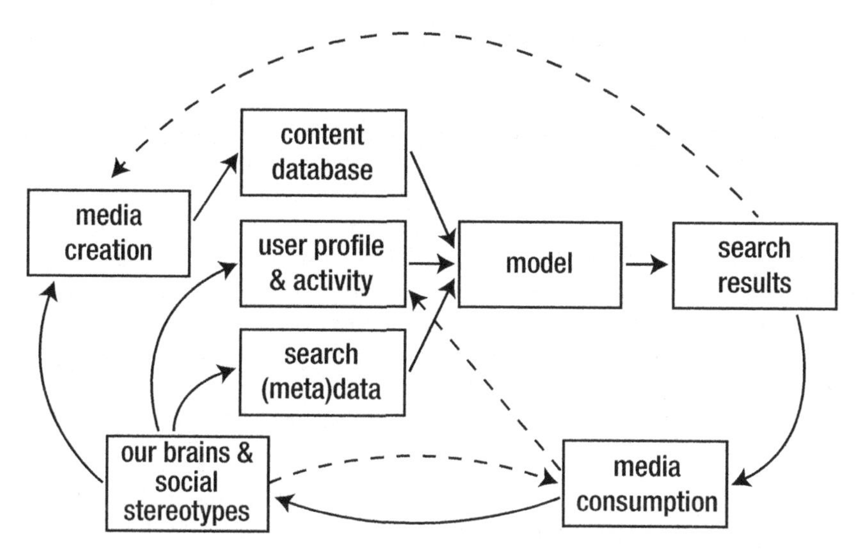


**Figure 3-1.** The feedback loop of human-algorithm interaction

But that's not the entire process. When we perform a search or we open up our Facebook or Twitter app, we do something with that algorithmically selected content. Consider the image searches from the Introduction—*doctor, nurse, professor, teacher*, etc.—and let's assume I want to create a web page for a health-related event. I search for stock images of doctors and nurses to include in that brochure. When I search for an image of a doctor, the search results will be filtered according to what the search knows (and guesses) about me, what it knows (and guesses) about users assessed to have similar tastes, what content it has in its database, and what general information it knows about that content and general engagement with it. As we know from the Introduction, the biases about what a doctor does/should look like that are present in the world will influence the search results, and the results will in turn influence our perception of the world and our biases, which will influence further search results, etc.

But we can add some nuance to that understanding. First, because of the processes of collaborative filtering, the biases I already experience *and the biases of people already similar to me* are the ones that will most strongly influence the output for my search. This is most starkly seen in Dylann Roof's alleged search for "black on white crime." Any objective crime statistics that might have offered at least a small check on his hateful extremism was masked by the fact that searches for the specific phrase "black on white crime" from users with similar internet usage patterns to Roof's were likely to filter out the more objective and moderate content from his results.

Bias amplification can be even stronger on social media platforms like Facebook or Twitter. There the content of a user's feed is already filtered by the people they are friends with and the pages they "like." Since we are already prone to associate more with people like us in some way than those who are not, that already represents a significant potential filter bubble. When our past engagement data and the results of the collaborative filtering process are also taken into account, the content we see can be extremely narrow. Intervening by following pages and befriending people who represent a wider range of perspectives can only help so much, as it affects the first filter, but not the collaborative engagement-based filter. This is why close friends or family members who have many friends in common may still see radically different content in their feeds. And since both the networks of friends/pages/groups we have curated and the posts we "like" and otherwise engage with tend to reflect our personal biases and limits of perspective, the content we encounter on the platform will tend to reflect those biases and limited perspectives as well.

That leads to a second point: if the content served up to me by algorithmic recommendation is biased in ways that reflect how I already think about the world, I am not only more likely to engage with that bias, *I am more likely to spread it*. An increasing number of people are finding their news on social media platforms.[14] But if it's easier to *find* information in a one-stop shop like Facebook or Twitter, just think of how much easier it is to *share* information found on that platform. With just a tap or two, I can repropagate an article, photo, or video I encounter—without necessarily even *reading* the article or watching the entire video, if the previewed content gets me excited enough. And this is true for algorithmic feeds like Twitter and Facebook in a way that isn't true for expertly curated content like that found in a print newspaper or a college textbook.

This sharing optimization compounds the filter bubble effect. Because it is easier to find information that reflects my existing biases *and easier to share it*, my contributions to others' social feeds will reflect my biases even more than if I only shared content that I found elsewhere on the internet. And, of course, the same is true for their contributions to my feed. This creates a feedback loop of bias amplification: I see things in accordance with my bias, I share a subset of that content that is chosen in accordance with that bias, and that feeds into the biased content the people in my network consume, from which they choose a subset in accordance with their bias to share with me, and so on. Just like the image search results in the Introduction (but likely more extreme), left unchecked this feedback loop will continue to amplify the biases already present among users, and the process will accelerate the more people find their news via social media feeds and the more targeted the algorithm

[14]Kevin Curry, "More and more people get their news via social media. Is that good or bad?," Monkey Cage, *The Washington Post*, published September 30, 2016, `www.washingtonpost.com/news/monkey-cage/wp/2016/09/30/more-and-more-people-get-their-news-via-social-media-is-that-good-or-bad/`.

becomes. And given the way that phenomena like clickbait can dominate our attention, not only will the things that reflect our own bias propagate faster in an algorithmically driven content stream, but so will content engineered to manipulate our attention. Put together, clickbait that confirms our preexisting biases should propagate at disproportionally high speeds. And that, in fact, is what we've seen happen in critical times like the lead up to the 2016 U.S. presidential election.[15] But, perhaps most importantly, the skewed media consumption that results will feed into our personal and social stereotypes about the world, influencing our behavior and relationships both online and in person.

Sometimes bias amplification works one-way. In cases like gender, racial, and other demographic representation we considered in the Introduction, the dominant group has been dominant long enough that algorithms tend to amplify the pervasive bias in that same, singular direction. But this is not always the case. When it comes to politics where, especially in the United States, we are relatively equally divided into two groups, the amplification of bias is not one-sided, but two-sided or multisided. The result, then, is *polarization*.

Polarization is easy enough to grasp. It simply means the increase in ideological difference and/or animosity between two or more opposing groups.[16] Digital polarization is in large part a result of the bias-amplification feedback loop applied to already like-minded groups. As biases get amplified within a group, it becomes more and more of a "filter bubble" or "echo chamber," where content uncritically promotes in-group thinking and uncritically vilifies the opposition. Adding fuel to the fire, we also know (as discussed in Chapter 2) that engagement increases when content is emotionally evocative—both positive and negative, and especially anger. This means not only more content supporting your view and discounting others, but more content is shared and reshared that encourages anger toward those outside the group. This makes it harder to listen to the other side even when more diverse content does make it through the filter.

## Letting Your Guard Down

There's another major problem that, left unchecked, causes algorithmically selected content to increase bias, polarization, and even the proliferation of "fake news." *Social platforms are designed for optimal engagement and primed for*

[15]Craig Silverman, "This Analysis Shows How Viral Fake Election News Stories Outperformed Real News On Facebook," *BuzzFeed News*, published November 16, 2016, www.buzzfeednews.com/article/craigsilverman/viral-fake-election-news-outperformed-real-news-on-facebook.
[16]"Political Polarization in the American Public," Pew Research Center, published June 12, 2014, www.people-press.org/2014/06/12/political-polarization-in-the-american-public/.

*believability*. Or, in the words of Renee DiResta, "Our political conversations are happening on an infrastructure built for viral advertising, and we are only beginning to adapt."[17]

There are several facets to this. First, social media encourages a relaxed posture toward information consumption and evaluation. By putting important news and policy debates alongside cat GIFs, baby pictures, commercial advertisements, and party invitations, social media puts us in a very different—and less critical—posture than a book, newspaper, or even magazine. Many users check social media when they are waiting in line at the store, riding the bus or train to work, even lying in bed. The relaxed posture can be great for social interactions, but that inhibition relaxation combined with the "attentional blink" that comes from shifting between cute cats and neo-Nazi counterprotests can make it difficult to think critically about what we believe and what we share.

Second, social media platforms are designed to promote engagement, even to the point of addiction. The change from a star to a heart for a Twitter "favorite," the increased emotion-based engagements on Facebook, all of these measures were taken in order to increase user engagement. And they worked.[18] Meanwhile, former employees of Google, Twitter, and Facebook have gone public with ways the platforms have been designed to promote addictive behavior,[19] and an increasing number of Silicon Valley tech employees have announced that they severely limit—or even prohibit—screen time for their own children.[20] Promoting engagement, even addictive behavior, alongside relaxed posturing is mostly harmless when it comes to baby pictures and cute cat videos, but it is not a recipe for careful, critical thinking around the major issues of the day.

Adding fuel to this fire, a recent study suggests that people judge the veracity of content on social media not by the *source* of the content but by the credibility of the *person who shared it*.[21] This means that even when we exercise

---

[17]Renee DiResta, "Free Speech in the Age of Algorithmic Microphones," WIRED, published October 12, 2018, www.wired.com/story/facebook-domestic-disinformation-algorithmic-megaphones/.

[18]Drew Olanoff, "Twitter Sees 6% Increase In 'Like' Activity After First Week Of Hearts," *TechCrunch*, published November 10, 2015, https://techcrunch.com/2015/11/10/twitter-sees-6-increase-in-like-activity-after-first-week-of-hearts/.

[19]Bianca Bosker, "The Binge Breaker," *The Atlantic*, published November, 2016, www.theatlantic.com/magazine/archive/2016/11/the-binge-breaker/501122/.

[20]Nellie Bowles, "Silicon Valley Nannies Are Phone Police for Kids," *The New York Times*, published October 26, 2018, www.nytimes.com/2018/10/26/style/silicon-valley-nannies.html.

[21]Tom Rosenstiel, Jeff Sonderman, Kevin Loker, Jennifer Benz, David Sterrett, Dan Malato, Trevor Tompson, Liz Kantor, and Emily Swanson, "'Who shared it?': How Americans decide what news to trust on social media," *American Press Institute*, published March 20, 2017, www.americanpressinstitute.org/publications/reports/survey-research/trust-social-media/.

critical thinking, we might be thinking critically about the wrong things. And platforms aren't much help here. Most social platforms take steps to highlight the sharer (bold text, profile picture, placement in the upper left of the "card") and downplay the originating source (smaller, lighter text, placement in the lower left of the card)—assuming the domain hosting the article is the primary source of the narrative, anyway.[22] This only exacerbates the tendency to judge a message by the messenger instead of the source.

Psychologists Nicholas DiFonzo and Prashant Bordia study the spread of rumors online, and they have identified four primary factors that contribute to whether or not someone believes a new idea or claim of fact they encounter:

- The claim agrees with that person's existing attitudes (confirmation bias).
- The claim comes from a credible source (which on social media often means the person who shared it, not the actual origin of the claim).
- The claim has been encountered repeatedly (which contributes to *perceptual fluency*).
- The claim is not accompanied by a rebuttal.[23]

In many ways, social media is engineered to bolster these factors, even for false narratives. We've already seen how engineering for engagement via collaborative filtering exacerbates the problem of confirmation bias and how on social media we often evaluate the wrong source. We've also discussed at length how repetition, especially when unconscious, can erode our ability to critically evaluate a false narrative. Both Twitter's retweet feature and Facebook's resurfacing of a story in our feeds every time it receives a new comment facilitate this. After all, we wouldn't want to miss out on a lively discussion among our friends! And we know that rebuttals spread far more slowly than the viral lies they seek to correct.[24] Social platforms as they exist today really are built to foster a sense of believability, regardless of what the truth really is.

Taking all these facts together, the deck is stacked against us for critical information consumption. Social platforms are designed in ways that make

---

[22]Mike Caulfield, "Facebook Broke Democracy, but the Fix Is Harder Than People Realize," Hapgood (blog), published November 10, 2016, `https://hapgood.us/2016/11/10/facebook-broke-democracy-but-the-fix-is-harder-than-people-realize/`.
[23]Bordia, Prashant and Nicholas DiFonzo. 2017. *Rumor Psychology: Social and Organizational Approaches*. Washington, D.C.: American Psychological Association.
[24]Alexis Sobel Fitts, "We still don't know how to stop misinformation online," *Colombia Journalism Review*, published October 9, 2014, `https://archives.cjr.org/behind_the_news/corrections_dont_go_viral.php`.

critical consumption hard and the sharing of unsubstantiated "truthy" claims easy. At the same time, more of us are getting more of our news from our social feeds, and we haven't even talked about the role that social media plays in how print, radio, and television journalists find their scoops and frame their stories. Left unchecked, fake news, filter bubbles, and polarization will only get worse. And given the importance of an informed electorate to a democratic republic, the algorithmic news feed represents a real threat to democracy.

That said, platforms have made changes in both attitudes and algorithms over the past two years. They still have a ways to go, and new "hacks" to the system are emerging all the time, but it feels to those of us who research this problem like we are finally headed in the right direction. It's time to capitalize on that and press forward as hard as we can, before new tactics and waning public pressure usher in a new era of complacency.

## Summary

So far, we've unpacked a lot of "theory" behind disinformation and online propaganda. The limits of human cognition, along with the affordances of social platforms and content recommendation algorithms, make it easy for half-truths and outright lies to spread, for biases to amplify, and for ideologically motivated groups to drift further away from each other—and further from the possibility of meaningful dialog and compromise.

These problems aren't simply organic. They don't only happen by accident, as flawed but well-meaning people interact on platforms designed by other flawed but well-meaning people. Polarization, bias, and distrust-at-scale are also fostered by bad actors, including state actors, who take advantage of these flawed systems and our increasing reliance on them.

In the remainder of this book, we will walk through several case studies that lay out a number of specific disinformation campaigns, or *influence operations*, that starkly depict the dangers we face when we rely uncritically on a small number of algorithmically driven platforms for our information consumption and political exchanges. But embedded in some of these case studies—and the way the public, governments, and/or platforms have responded to them—is hope as well. There are personal habits, platform modifications, and governmental tactics that work, which help digital media, even social media, become (again?) a force for good in the world.

That's what we'll explore together in the second half of this book...

PART

II

# Case Studies

CHAPTER

4

# Domestic Disturbance

## Ferguson, GamerGate, and the Rise of the American Alt-Right

For two whole days after the shooting of Michael Brown, major protests and rallies were taking place in Ferguson, Missouri, with hardly a mention in the mainstream news media, or even on Facebook. Yet a major portion of the country knew about the protests, the major influencers, and the emerging movement known as Black Lives Matter. Through Twitter, protesters were able to coordinate their activities, avoid tear gas and LRADs (long-range acoustic devices), even get phone charging battery packs to movement leaders. That activity on Twitter (and Vine) brought the events directly to the public eye and eventually to American mainstream news media. This watershed movement made many aware of the limitations of mass news media and the power of participatory media to shape conversations and effect social change. But those haven't been the only effects.

Almost simultaneously with the Ferguson unrest, a group of online "trolls" (harassers, abusers, and other kinds of bad actors, typically operating in the open) emerged to fight a culture war against "social justice warriors" they saw as threats to their world of online gaming communities. A number of game designers and critics—three women in particular—found themselves the victims of disinformation, coordinated online harassment, and credible threats


K. Shaffer, *Data versus Democracy*,
https://doi.org/10.1007/978-1-4842-4540-8_4

on their safety and that of their friends and family. Social platforms and law enforcement were slow and ineffective in their responses, and many lives were drastically changed by these attacks. For many, GamerGate was the pivotal event that made the dark powers of social media—including Twitter "bots" (automated accounts) and "sockpuppets" (fake accounts, managed by anonymous humans in real time)—apparent to a larger population. But those dark powers weren't finished...

In this chapter, we'll explore how some of the same affordances of social media can be used to grow movements on all sides of the political spectrum—movements meant to build up and movements meant to tear down. Movements that would change the American political landscape drastically, and likely for the long haul.

# Crowd Control: How the Tweets of Ferguson Steered Mainstream Media and Public Awareness

When announcing that there would be no indictment of Darren Wilson after the shooting of Michael Brown, St. Louis prosecutor Bob McCulloch stated:

> The most significant challenge encountered in this investigation has been the 24-hour news cycle and its insatiable appetite for something, for anything to talk about. Following closely behind with the non-stop rumors on social media.[1]

Protest leader DeRay McKesson responded on Twitter:

> McCulloch takes another dig at social media. But were it not for Twitter, he would've convinced the world that we didn't exist. #Ferguson[2]

Social and independent media have become cornerstones of (anti-)social activist movements in the twenty-first century. Arab Spring, Ferguson, #GamerGate, #BlackLivesMatter, #BlueLivesMatter, #MeToo—all have relied on social media to accomplish their purposes. This was seen clearly in Ferguson, where emergent activist leaders DeRay McKesson (@deray), Johnetta Elzie (@Nettaaaaaaaa), and Tef Poe (@tefpoe) used Twitter to

[1]Dylan Byers, "Ferguson Prosecutor Blames the Media," *Politico*, published November 25, 2014, www.politico.com/blogs/media/2014/11/ferguson-prosecutor-blames-the-media-199249.html.

[2]DeRay McKesson, Twitter Post, November 24, 2014, 9:20 p.m., https://twitter.com/deray/status/537068182909882368.

organize protests around St. Louis and to warn activists of police presence and activity. At the same time, protesters used Twitter, Instagram, and Vine to document alleged police brutality and to "cover" events outside the police-imposed press pen. And many of the journalists victimized by the police[3] or brave enough to embed themselves with the protesters (outside said press pen) used Twitter, Vine, and live-streamed video to get news out in real time.

The decentralized nature of these social media platforms empowers groups and individuals to get messages out fast, bypassing mainstream media outlets. They also allow for, perhaps even require, a different kind of activist leadership. In Ferguson, this led to criticism from figures like Oprah Winfrey calling for more leadership from the movement[4]—in other words, a centralized voice. But as activist responses made clear (on Twitter, of course), the movement does have leadership—only leadership of a different kind.

Shaun King's response to Oprah was telling:

> I love @Oprah. Was an Oprah Scholar @Morehouse. Her quote suggests 1 thing... All she learned about Ferguson came from the nightly news.[5]

The accusation was not that Oprah was wrong but that Oprah had the wrong information. That she doesn't understand the media being used by the movement and the difference in how the movement organizes around those media.

Oprah was not alone. Though Ferguson was primarily an in-person protest, its reach, longevity, and impact were in large part facilitated by social media, in particular Twitter and Vine. Understanding how Ferguson played out, and the impact it had on the rising #BlackLivesMatter movement, requires understanding the social media landscape and how it was used by activists to coordinate action, spread their message, and steer the narrative.

## August 9, 2014: What Happened in Ferguson

I heard it first on Twitter. In 2014, that statement could refer to just about anything. It was far and away my most active year on Twitter. I used it in my research, my teaching (I even *made* my students tweet in class, and we

---

[3]Jon Swaine, "Michael Brown protests in Ferguson met with rubber bullets and teargas," *The Guardian*, published August 14, 2014, www.theguardian.com/world/2014/aug/14/ferguson-police-teargas-rubber-bullets-michael-brown.

[4]Sarah Muller, "Protesters slam Oprah for suggesting movement lacks leadership," MSNBC, updated January 5, 2015, www.msnbc.com/msnbc/protesters-slam-oprah-suggesting-movement-lacks-leadership.

[5]Shaun King, Twitter Post, https://twitter.com/ShaunKing/status/551109555040829440, deleted but cited in ibid.

interviewed visiting scholars via public Twitter chat), and for collaborating with other researchers around the globe. I had two computer monitors in my office, and one was often entirely devoted to TweetDeck, my Twitter client of choice. One of my academic colleagues said my setup looked like I was trading stocks, as several columns of constantly updating tweets flew down my screen.

But in this case, I'm referring to Ferguson. The shooting of Michael Brown on August 9, the several hours the police left his lifeless body on the street, the memorials and the vigils in the days that followed.

I heard it first on Twitter when a friend of mine from college who lived in St. Louis at the time was at one of those vigils and was greeted afterward by police in riot gear as the mourners left the church. I heard it first on Twitter when police established a press pen, forbidding journalists from exercising their first amendment rights to observe and report on the protests and the police response.[6] I heard it first on Twitter that journalists Wesley Lowery and Ryan Reilly were arrested in a McDonalds while they recharged their phone batteries.[7] I heard it first on Twitter when police fired tear gas at the crowd, which included infants and young children. I heard it first on Twitter when police used long-range acoustic devices (LRADs), psychological weapons that disorient crowds and can cause permanent hearing damage. I heard it first on Twitter that Amnesty International had sent observers to Ferguson, to observe and document potential human rights violations perpetrated by the police.[8]

For several days, I heard it all first on Twitter. In fact, for at least three days, I *only* heard it on Twitter. (And the Vice News livestream—which I heard about on Twitter.) I remember seeing some particularly unsettling events take place and asking my wife—who has never had a Twitter account—*can you believe what's going on in Ferguson?!* She replied something along the lines of: *What's Ferguson?*

At the time, I had Twitter, she had Facebook, and we didn't have a television in the house. We consumed news on social media or on mainstream news web sites. And while my Twitter feed was dominated by Ferguson, her Facebook feed (like many Americans') was dominated by the Ice Bucket Challenge and cute pictures of friends' kids. And for at least two days, it seemed the mainstream media had yet to hear about Ferguson, too.

---

[6]Noam Cohen, "U.S. Inquiry Sought in Police Treatment of Press at Ferguson Protests," *The New York Times*, published October 26, 2014, www.nytimes.com/2014/10/27/business/media/-us-inquiry-sought-in-police-treatment-of-press-at-ferguson-protests-.html.

[7]"Michael Brown protests in Ferguson met with rubber bullets and teargas."

[8]Tierney Sneed, "Amnesty International Blasts Handling of Mike Brown Shooting, Ferguson Protests," *U.S. News & World Report*, published October 24, 2014, www.usnews.com/news/articles/2014/10/24/amnesty-international-blasts-handling-of-mike-brown-shooting-ferguson-protests.

In addition to bringing into sharp relief the vices of police brutality, police militarization, and racial injustice, Ferguson also brought into sharp relief the differences between mainstream media and social media and between algorithmically driven social platforms (like Facebook) and the still-reverse-chronological-order Twitter timeline. While all three—mainstream media, algorithmic social media, and reverse-chronological social media—give certain people access to more information and/or a larger audience, each type of platform privileges different voices and different messages, and each platform privileges different kinds of group organizing.

## A Movement Emerges as "Leaderless" Activists Organize on Twitter

As I watched the events in Ferguson unfold from a distance, I observed the organizational structure emerge in real time. In the beginning, there was little organization to speak of. People who were physically present, and those who weren't, were sharing information—some true and some later debunked. Much of this content contained the hashtag #ferguson, which I followed in its own column on TweetDeck (alongside a column containing the tweets of journalists who had proven trustworthy, to me, in their reporting). The early content that I observed was raw—information, emotion, decentralization, no singular community voice.

However, the community present on the ground had needs, and these needs shaped their conversation, and it began to take on a structure. As police formed barricades, protesters communicated both police locations and their own locations to each other, so they could avoid arrest, escape the tear gas, or form larger in-person groups to support each other. When police brought out tear gas, LRADs, or started firing rubber bullets, protesters shared that information with each other. Some took that as a cue to avoid the area, others as a cue to swarm in with their cameras and document what most of the press was not. When people needed medical assistance and ambulances could not make it through the police line or a crowd of protesters, they used Twitter to get the word out to those who might help. And, of course, experienced protesters shared tips for dealing with the police tactics that were emerging—where to find gas masks, how to make makeshift ones, why you should rinse eyes with milk instead of water, where to get earplugs, etc.

Twitter gave every participant a platform, and the #ferguson hashtag gave those participants an audience. Twitter also allowed people to "participate" from a distance—offering words of solidarity, connecting with organizations in other Midwestern cities, and offering expert tactical advice. But as word of the unrest spread, Twitter also allowed people to interfere, both locally and from a distance. Many who were talking *about* Ferguson were using the same hashtag as those trying to organize *in* Ferguson. Not all those talking about Ferguson were being helpful, and not every message being retweeted was worth amplifying.

But the hashtag didn't just give individuals a broadcast platform, it gave people a way to find each other—a "hacked public space"[9] that functioned as a kind of digital community center. That led to real personal connections and conversations—a way of building community and a way of vetting for agitators. Rather quickly, several voices emerged that were (1) demonstrably present, (2) reliable, (3) accessible, and (4) had a large reach. When the hashtag became noisy, or there wasn't time for every individual in the community to vet every claim of fact, these voices emerged as leaders who could be relied upon. They weren't just the people who knew and published the best information, they were also reliable conduits—people to tag when sharing information, so they could amplify good and pressing information to the community.

One of these emergent leaders wasn't even from St. Louis. DeRay McKesson, a middle school administrator from Minnesota,[10] heard about the protest online and quickly made his way down to support the movement in person, mainly on the weekends. By virtue of his expertise with both protest and with Twitter, he quickly became an emergent leader of the movement. And a target of the police. In an interview with *The Atlantic*, he remarks on the emergent nature of the Ferguson protest leadership:

> Ferguson exists in a tradition of protest. But what is different about Ferguson, or what is important about Ferguson, is that the movement began with regular people. … There are structures that have formed as a result of protest, that are really powerful. It is just that you did not need those structures to begin protest. … Twitter allowed that to happen.[11]

McKesson doesn't go into specifics in this interview about what most of those structures are. But we've already noted several of them. First, as McKesson does stress in that interview, a movement can begin with anyone. Social media gives everyone a voice, and if it's heard by the right core group of people, it can rapidly be amplified to a national or international audience. The leadership of the movement is also emergent. It may end up being people with deep local roots—like St. Louis alderman, Antonio French, who was a major voice on

[9]Dorothy Kim, "The Rules of Twitter," *Hybrid Pedagogy*, published December 4, 2014, `http://hybridpedagogy.org/rules-twitter/`.
[10]Noam Berlatsky, "Hashtag Activism Isn't a Cop-Out," *The Atlantic*, January 7, 2015, `www.theatlantic.com/politics/archive/2015/01/not-just-hashtag-activism-why-social-media-matters-to-protestors/384215/`.
[11]Ibid.

Twitter and who was arrested for alleged unlawful assembly[12]—it may be people who come to Ferguson from elsewhere to join in and take up the charge—like McKesson—or it may be people who are already speaking to a national audience about racial injustice at the hands of the police—like Alicia Barza, the founder of #blacklivesmatter, a movement that predated Ferguson but which rose to much greater public visibility as a result of Ferguson.[13]

Perhaps most importantly, the structure of the movement was agile and responsive to the needs of the moment. The leaders and the communicative strategies that emerged were the result of specific needs in the face of unanticipated circumstances. The combination of hashtags and @-mentions, the bifurcation of public tweets and direct private messages, the use of text and embedded Vine videos and livestreams, even the sharing and charging of phones—all of these specific practices that emerged were born of the needs of the moment, in conjunction with the affordances and limitations of the technology. To many outside the movement—or simply lacking in Twitter literacy—it likely seemed chaotic and "unplanned," but given the community's fluency with the tools, it allowed them a flexibility they would not have otherwise had. In fact, without Twitter and Vine, it's likely there would not have been a movement to speak of. Or, in the words of McKesson, "Missouri would have convinced you that we did not exist if it were not for social media."[14]

## Who Decides What Stories Get Told?

For all of Twitter's ability to facilitate both community organizing and the proliferation of the message, its overall user base is still fairly small. According to Pew Research, less than one-fourth of American adults used Twitter in 2014.[15] And given the racial tinge to the unrest and the high degree of social media segregation in the United States,[16] it's no surprise that it took so much time for the news to hit the mainstream. In fact, there were over 1 million tweets with the #ferguson hashtag before the first mainstream news story

[12]"Michael Brown protests in Ferguson met with rubber bullets and teargas."
[13]Elle Hunt, "Alicia Garza on the Beauty and the Burden of Black Lives Matter," *The Guardian*, published September 2, 2016, www.theguardian.com/us-news/2016/sep/02/alicia-garza-on-the-beauty-and-the-burden-of-black-lives-matter.
[14]"Hashtag Activism Isn't a Cop-Out."
[15]"Social Media Fact Sheet," Pew Research Center, published February 5, 2018, www.pewinternet.org/fact-sheet/social-media/.
[16]Robert P. Jones, "Self-Segregation: Why It's So Hard for Whites to Understand Ferguson," *The Atlantic*, published August 21, 2014, www.theatlantic.com/national/archive/2014/08/self-segregation-why-its-hard-for-whites-to-understand-ferguson/378928/.

about the unrest.[17] It didn't help that Ferguson police were actively working to control and contain the story, relegating (most) journalists to a "press pen" where they were kept far from the physical confrontations and fed police press releases, arresting journalists from *The Washington Post* and *The Huffington Post* while they filmed police activity, and even firing rubber bullets at a camera crew from Al Jazeera.[18]

But ultimately, the mainstream media did hear the story—and not just the police's story, but the protester's story as well. The FBI heard the story, too, and investigated (and later censured) the Ferguson Police Department for their actions. Amnesty International sent observers who also called out human rights violations perpetrated against the protesters.

But what if the story hadn't broken through? What if it had stayed in Ferguson?

It wasn't just the national mainstream media that wasn't covering Ferguson from the start. Facebook was eerily silent, too. As discussed earlier, while Twitter's reverse-chronological-order feed was dominated by #ferguson for many Americans, Facebook's algorithmic feed was not, sometimes even for the same people.[19] Algorithms make "decisions," and what they decide impacts what we see.[20] For whatever reason, Facebook's algorithm in 2014 "decided" that the Ice Bucket Challenge was a more appropriate content topic to deliver to most of its users than the Ferguson unrest. It was only on Twitter, where a tweet or a retweet immediately put that content at the very top of every follower's feed, that the story had an opportunity to take off. This not only increased the likelihood of those messages being seen (relative to an algorithm that deemphasized them), it also made the platform useful for organizing, which ensured that the tweets (and retweets) from the heart of Ferguson kept coming.

Twitter doesn't work that way anymore. Not long after Ferguson, Twitter introduced "While you were away"—an algorithmic interjection into an otherwise reverse-chronological-order timeline.[21] And in 2018, Twitter made an algorithmic timeline the default. Though they gave users the option to switch to a reverse chronology of the accounts they follow, the apps frequently return to the algorithmic default.

---

[17]Conrad Hackett, Twitter Post, August 20, 2014, 5:59 p.m., twitter.com/conradhackett/status/502213347643625472.
[18]"Michael Brown protests in Ferguson met with rubber bullets and teargas."
[19]Ibid.
[20]Zeynep Tufekci, "Algorithmic Harms Beyond Facebook and Google: Emergent Challenges of Computational Agency," in *Social Media Studies*, ed. Duan Peng and Zhang Lei, p. 213, accessed from https://ctlj.colorado.edu/wp-content/uploads/2015/08/Tufekci-final.pdf.
[21]Alex Kantrowitz, "An Algorithmic Feed May Be Twitter's Last Remaining Card To Play," *BuzzFeed News*, published June 29, 2015, www.buzzfeednews.com/article/alexkantrowitz/an-algorithmic-feed-may-be-twitters-last-remaining-card-to-p.

The reason for the algorithm is simple. The algorithm increases engagement and makes the service easy for new users to catch on to.[22] Both of those things increase advertising income and help Twitter make a profit. But that raises an important question: What happens to the next Ferguson? Will we hear about it? Or did we already miss it?

In the end, Twitter was a net positive for the people of Ferguson and the movement that was born hybrid—the physical and the digital each being critical components of its nature. It facilitated communication and coordination, as well as the spread of a message far wider than would have been possible even five years earlier.

As good as that was for #ferguson and #blacklivesmatter, it wasn't good for everyone.

## You Can't Just Quit the Internet: How GamerGate Turned Social Media into a "Real-life" Weapon

In 2014, a group of activists began to emerge, organize, and spread their message nationally on Twitter. The battle they were fighting was not new, but some of the tactics were, as were the means by which they organized and the realization that they were both many and powerful. This loose coalition of tech-fluent activists found each other, recruited others, and resisted the advances of the enemy. The impacts of their tactics online had a noticeable impact on the "real" (i.e., physical) world.

While those statements are all true relative to the protesters of Ferguson, that's not actually what I'm talking about here. In 2014, another group emerged onto the national scene, also largely via Twitter. That group was GamerGate. Rather than fighting for social justice, this group fought a decidedly *anti*social battle against those they dubbed "social justice warriors," or SJWs. These GamerGaters—predominately young, white, straight, and male—decried the diversification of gaming, considering it an attack against the straight white male "minority." But it wasn't just a battle of words or ideas, it was a battle that involved threats of physical violence, threats so real that the recipients of those threats fled for their lives. And it didn't just stop with GamerGate. Out of that very group formed a new movement—the so-called alt-right. This movement not only played a pivotal role in shifting the tone—and probably some votes—in the 2016 U.S. presidential election, but their culture war led to a shooting at a Washington, D.C., pizza parlor in December 2016 and the death of social activist Heather Heyer in Charlottesville, Virginia, in August 2017.

---

[22]Ibid.

But before we get into the alt-right, first we need to unpack the events that led to its rise: GamerGate.

## Zoë Quinn and the Blog Post from Hell

In 2013, independent game developer, Zoë Quinn, released the game *Depression Quest*.[23] It wasn't your stereotypical game. Of course, that's kind of the point. Ever since the 1980s, the public perception of video games has been one dominated by role-playing games (e.g., *World of Warcraft*), first-person shooters (e.g., *Call of Duty*), and sports games. The emphasis on competition, conquest, and violence in these stereotypical games is very closely tied to the stereotypical *gamer*—a suburban, adolescent (straight, white) male. Now there's nothing inherently masculine about games, nor about competition, nor even violence. But in Western culture, particularly in the United States, competitiveness, leadership, and heroism (too often confused with violence from one of the "good guys") have unfortunately become intertwined with an idea of what masculinity looks like.

Quinn's game—and her image as a gamer—could hardly be more opposite. A queer-identifying woman, who also implanted a chip into her body so she could become a "cyborg," Quinn represents a very different image of a gamer.[24] Likewise, her game *Depression Quest* sought not to give players a rush of adrenaline while killing Nazis or protecting Earth from an alien invasion. It wasn't even meant to be fun. Instead, *Depression Quest* sought to raise awareness of what individuals go through when they experience depression, something Quinn herself has experienced.[25] Whether or not it is a "good" game (I've never played it myself), it will certainly go down in history as an important game—not only for what it demonstrates that a game could be, but because of the movement that built up around it.

I wish I were talking about a movement that supported and encouraged expanding public ideas of what a game(r) could be. Unfortunately, that's not the movement that followed.

Quinn had already faced some blowback from "traditional"/stereotypical gamers about her game and the press that it received.[26] A game that pushed the boundaries the way *Depression Quest* did, especially one created by a

---

[23]Simon Parkin, "Zoë Quinn's *Depression Quest*," *The New Yorker*, published September 9, 2014, `www.newyorker.com/tech/annals-of-technology/zoe-quinns-depression-quest`.

[24]Noreen Malone, "Zoë and the Trolls," *New York Magazine*, published July 24, 2017, `http://nymag.com/intelligencer/2017/07/zoe-quinn-surviving-gamergate.html`.

[25]Ibid.

[26]Kyle Wagner, "The Future of the Culture Wars Is Here, And It's Gamergate," *Deadspin*, published October 14, 2014, `https://deadspin.com/the-future-of-the-culture-wars-is-here-and-its-gamerga-1646145844`.

female gamer from the LGBTQ community, was bound to be fodder for thinkpieces, and that naturally led to some negative (as well as positive) responses. But in August 2014, something else happened.

Quinn and her boyfriend, Eron Gjoni, broke up. It didn't go well. Others, including Quinn herself, have written about the details,[27] so I won't bother with most of them. The key outcome was that Gjoni, in a bout of premeditated rage, published a blog post accusing Quinn of sleeping with a journalist at the gaming news web site, Kotaku, in exchange for a positive review of her game. That review never existed. Nonetheless, the combination of years of pent up resentment about the diversification of gaming and a blog post crafted specifically to take advantage of that rage led to an online explosion. Discussion of "the Zoë post" spawned its own subreddit,[28] *quinnspiracy*, and dominated 4chan—a site for super-nerds that birthed memes, rickrolling, "epic fail," and the Trump Train, as well as copious sexism, racism, and, occasionally, the coordinated online harassment campaign—and 8chan—a site for people who think 4chan is too "politically correct." (In the words of Ben Schreckinger, 8chan is the ISIS to 4chan's Al Qaeda.[29])

## Antisocial Media: When Domestic Psychological Abuse Tactics Scale Up

The discussion didn't just stop at discussion. Things quickly turned violent. When Gjoni's post went live, Quinn recounts receiving a message from a friend, "you just got helldumped something fierce."[30] Shortly thereafter, old text messages, intimate photos, and flat-out fabrications were all over the darker corners of the internet. She was doxxed—her personal information, or private "docs," were published online—and complete strangers were calling her at all hours of the day and night. Family members were doxxed, and compromising, intimate, and/or threatening media were sent to her family, including via the mail. Accounts she forgot she had were hacked. And while this kind of thing has happened to people—especially women—before,[31] this time it just wasn't stopping. Months later, in January 2015, Quinn wrote a blog

[27]Ibid.; "Zoë and the Trolls"; "Zoë Quinn's *Depression Quest*"; Zoë Quinn, *Crash Override* (New York: PublicAffairs, 2017).

[28]Reddit's name for a specific Reddit community, usually organized around a discussion topic or group identifier.

[29]Ben Schreckinger, "World War Meme," *Politico Magazine*, March/April 2017, `www.politico.com/magazine/story/2017/03/memes-4chan-trump-supporters-trolls-internet-214856`.

[30]*Crash Override*, p. 10.

[31]Kathy Sierra, "Why the Trolls Will Always Win," *WIRED*, published October 8, 2014, `www.wired.com/2014/10/trolls-will-always-win/`.

post about the ongoing attacks entitled "August Never Ends."[32] And in the introduction to her 2017 book, *Crash Override*, Quinn wrote, "Most relationships end in a breakup. Sometimes that breakup is so crazy that it becomes a horror story you tell your friends, family, and therapist. ... My breakup required the intervention of the United Nations."[33]

In retrospect, and even as it was unfolding, it was clear that this was bigger than *Depression Quest* or Zoë Quinn. *GamerGate*, as Adam Baldwin dubbed it, was an outburst aimed at Quinn—and game developer Brianna Wu and feminist game critic Anita Sarkeesian[34] and pretty much anyone who spoke up publicly on their behalf—but it had been festering for some time. It's no coincidence that GamerGate rose at the same time that the indie gaming industry was really taking off—a time when the stereotypical game and gamer were being "pushed out" by an increasingly diverse market of games developed with an increasingly diverse group of gamers in mind.[35]

As fascinating—and truly gut-wrenching—as the sociology of GamerGate is, it's the tactics and the follow-up that are important for a book like this. Several key tactical trends emerged during GamerGate and continued on into subsequent campaigns, threatening public discourse in the United States and elsewhere, and even the integrity of our elections.

Just as the Ferguson activists organized online, GamerGaters organized online. But there was a key difference. They weren't using a popular platform like Twitter to organize and recruit for an event that was primarily taking place on the streets. Instead, they used less well-known platforms like 4chan, 8chan, and the seedier subreddits to organize and recruit for operations that took place on more public social media, like Twitter. In other words, Twitter wasn't the organizing platform, it was the theater of battle. It was where they fired up their sockpuppets and bots to spread doxxes and lies about their targets, propaganda about their mission, and attacks directed specifically at their targets.

The use of automated accounts, or bots, to spread disinformation was also a distinctive hallmark of GamerGate. I remember the first time I tweeted about GamerGate, using the term itself. I almost immediately received several replies from accounts with no credible personally identifiable information.

[32]Zoë Quinn, "August Never Ends," Zoë Quinn (blog), January 11, 2015, `http://ohdeargodbees.tumblr.com/post/107838639074/august-never-ends`.
[33]*Crash Override*, p. 1
[34]Nick Wingfield, "Feminist Critics of Video Games Facing Threats in 'GamerGate' Campaign," *The New York Times*, published October 15, 2014, `www.nytimes.com/2014/10/16/technology/gamergate-women-video-game-threats-anita-sarkeesian.html`.
[35]"Zoë and the Trolls."

One was a post containing the farcical GamerGate catch phrase, "actually, GamerGate is about ethics in gaming journalism." Another was a link to a YouTube video critiquing the ethics of Kotaku. None were at all a response to the content of my tweet, rather they were clearly a scare tactic meant to intimidate any outsiders to the movement who dare speak up about it.

These bots would respond to a variety of kinds of tweets. Several times, including once as late as 2017, if I posted a tweet containing the word "doxxed," a bot with an oil painting of an orthodox Jewish man as a profile picture and an anti-Semitic slur as a profile name would immediately reply to tell me that "doxxed" was a misspelling of "doxed." (The past tense of dox is acceptable both with one or two Xs, and I happen to prefer two.) This automated intimidation was a technologically cheap but socially powerful way to silence some voices and exert disproportionate control over the public discourse around the movement.

## Unprepared: How Platforms, Police, and the Courts (Failed to) Respond

Despite the lies, the targeted harassment, and the physical threats of violence, it was difficult for the victims of GamerGate to receive help, both from platforms and from law enforcement. Law enforcement didn't know what to do with Quinn's police reports.[36] Brianna Wu unsuccessfully called for the federal government to investigate and prosecute the owner of 8chan.[37] Citing "free speech," Twitter was exceedingly hesitant to shut down accounts. And a judge in a criminal harassment case (the accused was acquitted) suggested that if Quinn wanted to avoid harassment, perhaps she should stay off the internet. When she reminded him that her work as an independent game developer required not only an online presence, but a public social media presence, he responded, "You're a smart kid. … Find a different career."[38]

While the attacks have never completely stopped, Quinn, Sarkeesian, and Wu have braved the fight and emerged as strong voices for change both in the gaming industry and at the social media platforms. Quinn and Sarkeesian spoke to the United Nations about online abuse and harassment.[39] Quinn founded a company called the Crash Override Network that helps individuals

[36] "August Never Ends."
[37] Briana Wu, "I'm Brianna Wu, And I'm Risking My Life Standing Up To Gamergate," *Bustle*, published February 11, 2015, www.bustle.com/articles/63466-im-brianna-wu-and-im-risking-my-life-standing-up-to-gamergate.
[38] "Zoë and the Trolls."
[39] *Crash Override*, p. 115.

targeted by coordinated online abuse work with the platforms to shut down those who break the law and the Terms of Service in the course of their attacks.[40] Wu even ran for U.S. Congress.[41] And all three of them continue to work in gaming, despite GamerGate's efforts to push them—and many other women, people of color, and members of the LGBTQ community—out.

But in addition to the public awareness that Quinn, Sarkeesian, Wu, and others have brought to the social problems at the root of GamerGate—and the ways that platforms like 4chan, 8chan, Reddit, and Twitter have enabled their campaigns—other key figures and movements emerged during GamerGate. GamerGaters themselves also discovered their power.

GamerGaters weren't just jerks on the internet. They were a group of tech-fluent individuals, many of whom had spent significant portions of their lives online, particularly on image boards like 4chan. The affordances and limitations of those forums, and the community practices that emerged in light of those platform structures, facilitated certain tactical strengths within the GamerGate community that they used beyond GamerGate itself. For instance, 4chan is a platform where threads of content are deleted relatively quickly. Similarly, Reddit's system of upvotes and downvotes, coupled with its highlighting of high-engagement content on the front page of the site, motivates and rewards the creation of viral content. 4chan veterans, in particular, often have a keen sense of what will go viral, or at the very least, evidence of what did and didn't go viral on their platform. This awareness of patterns of virality, often combined with an obsession over detail—what some 4chan users reprehensibly call "weaponized autism"[42]—and skills in Twitter automation, gives them an advantage when it comes to creating and spreading memes and other potentially viral content.

GamerGate also allowed several key figures to emerge as leaders of the movement, whose leadership persisted beyond the bounds of GamerGate itself. Two key figures in that regard are former Breitbart editor, Milo Yiannopoulos, and independent pundit/provocateur, Mike Cernovich.[43] Both used their social media prowess to keep the movement going (Yiannopoulos, was later banned from Twitter for his role in mobilizing a harassment campaign against *Ghostbusters* star, Leslie Jones,[44] and Mike Cernovich played a key role

[40]`www.crashoverridenetwork.com`.
[41]`www.briannawuforcongress.com`.
[42]"World War Meme."
[43]"Zoë and the Trolls."
[44]Joseph Bernstein, "Alt-White: How the Breitbart Machine Laundered Racist Hate," *BuzzFeed News*, Published October 5, 2017, `www.buzzfeednews.com/article/josephbernstein/heres-how-breitbart-and-milo-smuggled-white-nationalism`.

in spreading the #pizzagate conspiracy theory[45]). And both were able to use their credibility inside and outside the GamerGate community to build bridges between a variety of tech-fluent, antifeminist, far-right, and even extremist communities. Much like Twitter enabled bridges to be built between Ferguson protesters, the emerging Black Lives Matter movement, and a variety of other existing organizations and sympathizers, Twitter also enabled various antifeminist, and in some cases blatantly white nationalist, groups to coalesce into a loose, but connected, movement that has been dubbed the alt-right.

## The Emergence of the Alt-Right

There were several key social strands to #GamerGate. First was the opposition to feminism, specifically, and diversity, generally, in the world of gaming. Second was the group dynamic—the social organizing on the deeper, darker corners of the web for "ops" that took place on the more open web. There was also an uneasy juxtaposition of libertarian and far-right politics alongside a "seriously, who CARES?!?!" mentality. The former rallied around an ideology of "free speech" and antipolitical correctness, the latter joined in these ops "for the lulz" (i.e., LOLs).[46]

These trends didn't go away, even as GamerGate finally started to die down. Though in a very real sense, GamerGate never died down. It just shifted its focus from games to politics.

Several journalists and new media scholars have written about this transition. Author Dale Beran writes of 4chan in general that their "only real political statement" is that "all information was free now that we had the internet."[47] Regarding GamerGate in particular, social justice warriors were encroaching on that freedom by "adding unwanted elements into their video games, namely things that promoted gender equality," and that this was part of "a grand conspiracy perpetrated by a few activists to change video games." Some took it upon themselves to dig for information online that would uncover this conspiracy, and since "all information [is] free," it was perfectly acceptable—nay, it was their duty as citizens—to share and expose the information they uncovered.

---

[45]"Mike Cernovich," Southern Poverty Law Center, `www.splcenter.org/fighting-hate/extremist-files/individual/mike-cernovich`.
[46]"Why the Trolls Will Always Win."
[47]Dale Beran, "4chan: The Skeleton Key to the Rise of Trump," Dale Beran (blog), published February 14, 2017, `https://medium.com/@DaleBeran/4chan-the-skeleton-key-to-the-rise-of-trump-624e7cb798cb`.

This should sound familiar. The idea of a deep conspiracy carried out by a silent but powerful minority, directed at taking away the freedom of the majority under the banner of social justice—that is messaging that resonated with many during the 2016 presidential election in the United States (among other political contexts around the globe, some of which we will explore in the chapters to follow). And this "Deep State" conspiracy perpetrated by the "globalists" of the Democratic Party and their cronies throughout the world was one of the key underlying political theories of the movement that became known as the alt-right, as well as the foundational belief system for President Trump's first chief of staff, Steve Bannon.[48]

But the similarities between GamerGate and the alt-right are not simply ideological. The alt-right is made up of many of the same people as GamerGate. In a landmark piece of investigative journalism, Joe Bernstein's article "Alt-White: How the Breitbart Machine Laundered Racist Hate" chronicles many of the specific connections between these movements.[49] Specifically, Milo Yiannopoulos, who rose to prominence during GamerGate, worked closely with Steve Bannon and the Mercer family (major right-wing political donors who, among other things, substantially funded UK's Vote Leave campaign in 2016) while an editor at Breitbart and helped Bannon bring "the 4chan savants and GamerGate vets" into the fold. Bernstein's deep dive into a trove of Breitbart emails and other previously secret documents reveals connections fanning out from Yiannopoulos not just to Bannon and the Mercers, but white nationalist Richard Spencer, Devin Saucier (of American Renaissance, categorized as a hate web site by the Southern Poverty Law Center), Andrew "Weev" Auernheimer (administrator of neo-Nazi hate site, The Daily Stormer), indicted and ousted former Trump staffers Sebastian Gorka and Michael Flynn (through his son Michael Flynn, Jr.), and even *Duck Dynasty*'s Phil Robertson. While the top brass at Breitbart presented a less explicitly racist version of the alt-right, Bernstein's investigation makes it clear that actual neo-Nazis—like those who organized the Unite the Right rally in Charlottesville, Va.—were just a forwarded email away from Bannon and others who worked at the highest levels of the Trump White House.

More important for our considerations, though, is that the *tactics* of GamerGate continued into the alt-right. In an article for Politico, "World War Meme," Ben Schreckinger writes about how the culture of 4chan and the tactics of GamerGate made their way into the 2016 U.S. presidential election.[50] In fact, in many ways, we can see Hillary Clinton as simply the highest profile GamerGate victim.

---

[48]Daniel Benjamin and Steven Simon, "Why Steve Bannon Wants You to Believe in the Deep State," *Politico Magazine*, published March 21, 2017, `www.politico.com/magazine/story/2017/03/steve-bannon-deep-state-214935`.
[49]"Alt-White: How the Breitbart Machine Laundered Racist Hate."
[50]"World War Meme."

Schreckinger writes that "the white nationalist alt-right was forged in the crucible of 4chan," a community that was "preoccupied with gender politics." Some 4chan members saw themselves at war with the left. Others saw electing someone like Donald Trump to the presidency as a massive prank, or "cosmic joke." For the lulz, indeed.

Just like GamerGate, these 4chan vets used 4chan's "/pol/" board (short for "politically incorrect") as a "staging ground." Like in GamerGate, their fluency with a platform that was originally an image board and which requires constant engagement to keep a thread from getting deleted meant that they knew a thing or two about creating and propagating viral media. They used this to their advantage.

Much of what was workshopped in 4chan was seeded or market tested on Reddit. Even in most of the less savory subreddits, users tend to be more mainstream than those on 4chan. The cream of the crop on 4chan often crossed over to Reddit, where the system of up- and downvotes helped filter for the most viral-ready content. (Though Schreckinger claims that alt-righters "juiced the rules" to promote some of their content, anyway.[51])

The content that proved the most successful on 4chan and Reddit—the most accessible to "normies"—was then seeded on Twitter and Facebook, where it was easy for them to spin up the same kinds of automated (bot) and sockpuppet accounts that they did during GamerGate.

But there was one more wrinkle. The alt-right wasn't entirely a grassroots movement of like-minded individuals who found each other and organized online. There were also well-funded political operatives who had observed the power of this community during GamerGate and were willing to put them to work in electoral politics. According to Schreckinger, early on in the election, the Trump campaign was monitoring Reddit, in particular one subreddit named The_Donald. Schreckinger describes The_Donald as "a conduit between 4chan and the mainstream web," and numerous memes and videos that originated on 4chan made it through The_Donald into the mainstream—including some that were shared by Trump campaign staffers and even Donald Trump himself. And this conduit was allegedly the strongest during the time that Steve Bannon was in charge of Trump's campaign.

---

[51]Ibid.

## The Mob Rules -or- Who Decides What Stories Get Told? [redux]

Ultimately, we'll never know the full impact that social platforms had on these domestic U.S. social issues in 2014 and beyond. We won't know how many lives were saved (or lost) by the increased public awareness and mobilization around police violence and racial discord. We'll never know how the lives of the victims of GamerGate would have been different if the social web hadn't provided their attackers the platforms for finding each other and microtargeting their messages of hate and violence. We'll never know how many votes were changed or suppressed by influence operations leading up to the 2016 election.

Of course, there are some things we *do* know. We know that police violence and racial tension did not begin, or end, in Ferguson. We know that prominent women—especially those who rock the boat or speak out about injustices—have always been targeted more than men in the same position, especially in the world of tech. We know that the influence operations perpetrated by alt-right meme warriors (as well as some on the left, and—as we'll explore in the next chapter—foreign agents seeking to interfere in U.S. politics) changed the tone of the election and influenced the topics we were debating.

And we know that social media played a significant role in shaping how all of these events turned out. As DeRay McKesson said, those of us outside the St. Louis area might not have ever known about Ferguson—let alone joined in or raised our collective voices—were it not for Twitter. On the other hand, the same openness that allowed activists to organize in Ferguson allowed attackers to organize and perpetrate their attacks against women and their allies during GamerGate. But from yet another perspective, that openness is what gave voice, reach, and new business opportunity to independent game developers like Quinn and Wu and critics like Sarkeesian. And for better or for worse, the affordances of forums like 4chan, with its disappearing threads, and Reddit, with its upvotes and downvotes, honed the skills of those who sought to create and propagate viral media beyond those platforms.

To some extent, we will always have to take the good with the bad when it comes to technology. But that doesn't mean that the tools—or the people and companies who create and maintain those tools—are neutral. And it certainly doesn't mean that they don't bear at least *some* responsibility for what happens on their platforms, even if in some cases that responsibility is moral and ethical rather than explicitly legal.

When a judge tells an independent online game developer to stay off the internet and find a new career if she wants to avoid harassment and abuse, that's akin to telling someone to avoid the mall or the grocery store.

The internet is a core part of our lives in the twenty-first century, and social media is a "mediated public space," where those who own and run the "private" platforms have a responsibility to at least do their due diligence toward providing safe access for those they invite to make use of those platforms. But we don't need to wait for them to do it. Like during many technological revolutions in past centuries, it will take the will and the work of the people at large to shape the evolution of these "mediated public spaces" and to advocate for legislative change where necessary to ensure fair, safe access for all.

To that end, I leave you with two inspirational quotes from two prominent women in tech.

> People believe they are powerless and alone, but the only thing that keeps people powerless and alone is that same belief. People, working together, are immensely and terrifyingly powerful.
>
> —Quinn Norton[52]

> Although what was done to me wavs heinous, those responsible for obliterating my old life have overlooked one important thing: I'm better at games than they are.
>
> —Zoë Quinn[53]

# Summary

In this chapter, we explored three distinct but related historical events where the affordances and limitations of social media platforms had a significant impact on the way those events played out: the protests following the shooting of Michael Brown in Ferguson, Missouri, the GamerGate attacks on prominent women in the video game industry, and the mobilization of the alt-right during the 2016 U.S. presidential election. In each case, the affordances of platforms like Twitter, 4chan, and Reddit—combined with the ideology and customs of the communities that inhabited them—led to particular kinds of novel "operations." Social media is a powerful tool for community organizing and for recruiting new members to a movement. But it is not always used for social good, and the limitations of these platforms and the online space in general

---

[52]Quinn Norton, "Everything Is Broken," *The Message*, published May 20, 2014, `https://medium.com/message/everything-is-broken-81e5f33a24e1`.
[53]*Crash Override*, p. 7

can pose challenges to law enforcement, and on the grander scale to legislators and platform administrators, as they seek to balance the rights of free speech and free assembly with the rights to life, liberty, and safety.

In many ways, 2014 was a beginning, or at least a point of major acceleration, when it comes to social media's impact on society. And it wasn't just in the United States. In the chapters that follow, we'll unpack operations outside those borders, including operations that cross borders. Operations that involve far more than serendipitous community organizing.

CHAPTER

5

# Democracy Hacked, Part 1

## Russian Interference and the New Cold War

Social media empowers communities of activists, as well as groups of extremists and abusers, to discover each other and coordinate their activity. The same tools can be used to spread political messages, both by legitimate communities and by disingenuous actors—even foreign states seeking to interfere in the electoral process of another country. That's the environment we find ourselves in today, as the United States, NATO, the EU, and their (potential) allies are under attack from a Russian campaign of information warfare. In this chapter, we'll unpack some of their operations, culminating in the 2016 U.S. presidential election, and conclude with a view toward future threats and defenses.

## What Happened?

November 8, 2018. Hundreds of millions of Americans—and many others throughout the world—watched the results of the U.S. presidential election roll in. The odds and the polls all pointed to a Hillary Clinton victory. The question was simply how big a victory, and whether her party would control Congress as well.


K. Shaffer, *Data versus Democracy*,
https://doi.org/10.1007/978-1-4842-4540-8_5

Of course, that's not how things turned out. Even though Clinton won the popular vote, Donald Trump won victories in key swing states, giving him the edge in the Electoral College to become the 45th president of the United States.

Almost immediately after he was declared the presumptive winner by the major news networks, pundits and scholars began to ask, "What happened?" Even some who supported Trump over Clinton were surprised at his victory. Fingers quickly pointed at pollsters and their methods, which are still being updated for the age of the internet and smartphones (as opposed to one-per-family landlines). They also pointed to perceived biases in the mainstream media, reporting from one perspective and blind to others. Many voters disaffected with the two-party system and the major parties' nominees stayed home or voted for a third-party candidate, winning them the ire of some Democrats. And, as is frequently the case when the winning candidate loses the popular vote, there were calls for the abolition of the Electoral College.

But to many, the blame lay primarily at the feet of Hillary Clinton herself. Not only would old scandals simply not go away—Benghazi, her response to allegations of sexual impropriety against her husband Bill when he was president, her "socialist" tendencies as a past supporter of universal health care—but new ones kept popping up. Her use of a private email server for official State Department business suggested that she was a security risk. The missing emails from that server surely must have contained classified information (which would be illegal) or reference to other nefarious activities. Private communications revealed that the Democratic National Committee took steps to ensure that the more moderate Clinton won the party's nomination over democratic socialist Bernie Sanders. Three weeks before the election, then FBI Director James Comey announced that the Department of Justice was reopening the case of Clinton's private email server. And that's to say nothing of the more fringe rumors that Clinton had ordered murders be committed on her behalf,[1] was involved in "spirit cooking,"[2] or sat at the center of a massive child sex trafficking ring[3]—the latter of which, though baseless, led a man to bring a gun to a D.C. pizza parlor to "investigate." (None of these conspiracy theories have completely gone away.) As a result,

---

[1] "FBI Agent Suspected in Hillary Email Leaks Found Dead in Apparent Murder-Suicide," David Mikkelson, *Snopes*, accessed January 3, 2019, www.snopes.com/fact-check/fbi-agent-murder-suicide/.

[2] "Was Clinton Campaign Chairman John Podesta Involved in Satanic 'Spirit Cooking'?," Dan Evon, *Snopes*, published November 4, 2016, www.snopes.com/fact-check/john-podesta-spirit-cooking/.

[3] "Anatomy of a Fake News Scandal," Amanda Robb, *Rolling Stone*, published November 16, 2017, www.rollingstone.com/politics/politics-news/anatomy-of-a-fake-news-scandal-125877/.

many moderates and Sanders supporters who otherwise would have supported Clinton over Trump voted third party or abstained in protest, at least partially contributing to Trump's victory.

While some of these scandals had at least a kernel of truth to them and most had been shared in earnest by many Americans, it became clearer and clearer throughout 2017 that there were other forces at work. Yes, Clinton appeared mired in scandal, and yes Trump seemed to represent people and ideas that many "mainstream" journalists and political operatives paid too little attention to. But, we now know, not all of that was organic. Someone had hacked the DNC server, stolen the emails of Clinton's campaign chair, John Podesta, and boosted both the scandals they contained and the pro-Trump movements on social media. Many of the conspiracy theories were circulated and amplified, if not created, by disingenuous operatives running fake accounts on a variety of social media platforms. As researchers and the American intelligence community now agree, all signs point to one primary culprit: Russia.

Unfortunately, the same lack of nuance that led many to fall prey to Russia's activities—and to other influence operations that took place at the same time—has led many to either make Russia a boogeyman, responsible for all narratives and political victories they dislike, or to cheapen the propaganda problem as "Russian [Twitter] bots" and laugh it off. The former causes us to overlook both other foreign actors and the role that domestic bad actors and regular citizens play in the spread of disinformation. The latter causes us to overlook both the breadth and the gravity of the problem altogether.

# Meet the New War, Same as the (C)old War

The reality is that Russia describes itself as being in a state of "information warfare against the United States of America" as well as our NATO allies.[4] For years, Russia has been using web sites, blogs, and social media as the latest tools in their information arsenal against the United States, the United Kingdom, the EU, NATO, Ukraine, and Syrian rebels. Their goals are to enlarge their "empire,"[5] enrich Putin's "inner circle," weaken the EU and NATO, discredit Western democracy, and—at least in the immediate regions surrounding Russia—promote an illiberal faux democracy, where a few corrupt oligarchs can manage a state economy in their own favor.[6]

---

[4]Martin Kragh and Sebastian Åsberg, "Russia's strategy for influence through public diplomacy and active measures: the Swedish case," *Journal of Strategic Studies* 40/6 (2017), DOI: 10.1080/01402390.2016.1273830, p. 6.

[5]Stephen Blank, "Moscow's Competitive Strategy," American Foreign Policy Council, published July 2018, p. 2.

[6]Heather A. Conley, James Mina, Ruslan Stefanov, and Martin Vladimirov, *The Kremlin Playbook* (Lanham: Rowman & Littlefield, 2016), p. 1ff.

But Russia is not the only player. The far-right, populist groups that Russia has bolstered in the United States and Europe are real. The refugee crisis triggering the anger and activism of those anti-immigrant populists is also real—though Russia's activity in Syria and elsewhere certainly fuel that crisis. These groups were already having an impact within their own countries before Russia stepped in and lent them a hand—if they even needed to—and they continue to have a real influence, irrespective of any help they might get from Russia or their allies.

And there are other nations using online propaganda to further their geopolitical aims. Seven authoritarian nations have a budget for influence operations and propaganda,[7] and Twitter recently announced discovery of an Iranian operation aimed at Western users of their platform. Private companies from the United States, the United Kingdom, and Israel are certainly major players. And the GamerGate trolls haven't gone away, either.

In what follows, we'll survey this new theater of battle, getting the lay of the land in this global information war. Who are the players, what battles have been fought, how have platforms and governments responded, and how we got to where we ended up in 2016, where Russian influence operations played a significant, and possibly decisive, role in the election of the president of the United States.

## Ukrainian "Separatists"

There's no better place to start exploring Russian online disinformation than Ukraine. According to an exhaustive RAND study on Russian influence operations in Ukraine, "The annexation of Crimea in 2014 kicked off the debut of online Russian propaganda on the world stage."[8] Coordinated with military operations in Ukraine and worldwide diplomatic operations aimed at global recognition of the invasion, Russia conducted information warfare both within Ukraine and without, in order to reinforce the idea of Crimea's "Russianness" to Crimeans, Ukrainians, and the larger global community.

The RAND study found several important strains in Russia's influence operations around the Crimean invasion. First, they targeted ethnic Russians in Crimea and eastern Ukraine, many of whom were Soviet-era transplants

[7]Philip Howard in: "Foreign Influence on Social Media Platforms: Perspectives from Third-Party Social Media Experts," U.S. Senate Select Committee on Intelligence, Open Hearing, August 1, 2018, www.intelligence.senate.gov/hearings/open-hearing-foreign-influence-operations'-use-social-media-platforms-third-party-expert.
[8]Todd C. Helmus, Elizabeth Bodine-Baron, Andrew Radin, Madeline Magnuson, Joshua Mendelsohn, William Marcellino, Andriy Bega, and Zev Winkelman, Russian Social Media Influence: Understanding Russian Propaganda in Eastern Europe (Santa Monica: RAND Corporation, 2018), DOI: 10.7249/RR2237, p. 15.

and their descendants. Russian-language media still dominates the Ukrainian information landscape, and they took advantage of this inroad to promote unity between Russia and ethnically Russian Ukrainians. This had the triple advantage of promoting the Russianness of Crimea, encouraging pro-Russia political activity within Ukraine more broadly, and supporting pro-Russia Ukrainian separatist movements (in conjunction with military support of those same separatists). Ukraine has responded, in part, by blocking Russian media access in 2014, but as of 2017, the Russian social media platform VKontakte was still the most popular in Ukraine, followed by American-owned Facebook.[9]

Russia also targeted Ukrainians with its now standard Euro-skeptic narratives, lest the EU and/or NATO expand to encompass one of Russia's chief trade partners and line itself up on Russia's southwestern border. Paired with this Euro-skepticism came the narratives of corruption around Ukrainian political leadership. Russian outlets often portrayed Ukraine as being governed by white supremacist fascists.[10] Russia, of course, was Ukraine's stable neighbor and friend, stalwart of traditional moral, Christian values.

In the "far abroad"—more distant members of the world community, not sharing a border with Russia or one of its historical Warsaw Pact vassals—Russia advanced narratives in a variety of languages echoing many of the pro-Russia and anti-Ukraine messages it shared more locally. While the goal locally was to bring Ukraine back into Russia's sphere of influence—if not its direct control—the goal globally was primarily one of non-intervention. By advancing narratives of Ukrainian corruption, Russian virtue, Crimea's historical and ethnic Russianness, promoting fear of escalated global conflict, and sowing distrust of the United States, EU, and NATO, Russia encouraged the global community to accept Crimea's annexation, or at least not to intervene.

Perhaps the starkest example of Russia's anti-Ukrainian influence operations involves the false narratives around the tragedy that occurred to Malaysia Airlines Flight 17 in July 2014. While Western intelligence and the Dutch-led Joint Investigation Team have concluded that Russia and/or Russian-backed Ukrainian separatists are to blame,[11] Russia used the tragedy to smear Ukrainian forces as immoral and incompetent and Russia as an important stabilizing force. Along the way, though, multiple contradictory explanations were advanced by Russian propaganda outlets. Because, for all the value of

[9]Mariia Zhdanova and Dariya Orlova, "Ukraine: External Threats and Internal Challenges," in *Computational Propaganda*, ed. Samuel Woolley and Philip N. Howard (Oxford: Oxford University Press, 2018), 47.
[10]Russian Social Media Influence, p. 104.
[11]Zhdanova and Orlova, p. 55.

people believing the anti-Ukrainian narrative, Russia had to know they would be found out. However, by advancing multiple narratives through different outlets to different audiences, they sow confusion, give the investigators more to investigate, and prolong the general sense of disorientation. And for many, that sense of "I don't know whom or what to believe"—"paralysis through propaganda"[12]—will last longer than knowledge of the ultimate finding of investigators, especially if those findings are delayed by Russian obfuscation until new stories dominate the news cycle.

It's important to note that, for all the talk of bots in other locales, bots appear to have played only small roles in Ukraine. According to Ukrainian media and propaganda researchers Mariia Zhdanova and Dariya Orlova, "automated bots seem to be less widespread in Ukraine." Instead, "manually maintained fake accounts are one of the most popular instruments for online campaigns" aimed at the Ukrainian public.[13] This may be due to the relatively low saturation in Ukraine of bot-friendly platforms like Twitter, when compared to platforms like Facebook, where automation is more difficult and therefore more expensive.

Though the annexation happened in 2014, and global attention has largely been turned elsewhere, Russia's information warfare around Ukraine is ongoing, even in English. In my research into Russian disinformation, I regularly encounter new (to me) web sites and Twitter and Facebook accounts advancing anti-Ukraine narratives and promoting the right to self-rule of the "separatists" in Donetsk, Luhansk, and Novorossiya. Once these assets are developed, it is rare that Russia removes them voluntarily—though they may retask them. Rather, the inexpensive, and at times effective, operations continue on, until the asset is compromised or deleted by the platform on which it operates.

## Active Measures in the Baltic

Sweden, along with Finland, has long played a significant role as a buffer between Russia (and formerly the USSR and Warsaw Pact countries) and NATO. Because of this, Soviet spies were incredibly active in Sweden during the Cold War, gathering intelligence and conducting operations, in the hopes of preventing any eastward NATO expansion.[14] In the twenty-first century, the threat to Russia is less military and likely more economic (despite Russian statements to the contrary). To get exports, like oil, out of the Baltic Sea, Russia has to pass between Denmark (NATO and EU member) and Sweden.

[12]Russian Social Media Influence, p. 9.
[13]Zhdanova and Orlova, p. 51.
[14]Kragh & Åsberg, p. 8.

Given the economic pressure, including sanctions, that the United States and NATO have put on Russia's oligarchy since 2009—and the even stricter sanctions placed on their geopolitical ally, Iran—the threat of NATO encroachment via Sweden is real: not the threat of a military invasion, but the threat of a trade blockade in the Baltic, enforced by or through Sweden.

So when Sweden started considering in 2015 an agreement with NATO that would allow NATO military forces access to Swedish territories, Russia acted. In the information space.

In early 2015, state-run propaganda outlet, Sputnik News, launched a Swedish-language web site. Through that web site, a television network run by RT (formerly "Russia Today," an international, Russian-state-owned media outlet), and numerous covert channels, Russia flooded Sweden with propaganda.[15] This propaganda included standard Russian narratives: anti-NATO messaging, fear mongering about impending nuclear war (caused, of course, by U.S.-led NATO action), narratives about how the EU is in decline, and even anti-GMO and anti-immigration narratives.[16] This was, of course, alongside the dissemination and amplification of false narratives surrounding the Sweden-NATO agreement.

Researchers Martin Kragh and Sebastian Åsberg studied this Russian influence operation extensively. They found that, like is common for Russia, "Misleading half-truths are the norm" and "outright fabrications occur on a limited scope."[17] While they did find a number of easily disproven fabrications and forgeries, most Russian messaging amplified Russia-friendly narratives that already existed in Swedish culture, particularly from groups on the fringe.

One sequence of false narratives stands out, in particular. Beginning in late 2014, there were several sightings of unidentified foreign submarines on or near the Swedish coast. These sightings were credible and were reported in legitimate, mainstream outlets. But these weren't the first time that strange things had happened involving foreign submarines on the Swedish coast. Kragh and Åsberg write:

> When the Soviet submarine S-363 ran aground in 1981 on the south coast of Sweden, a forged telegram soon appeared in media purportedly written by the Swedish ambassador to Washington, Wilhelm Wachtmeister. The telegram expresses the ambassador's

[15]Neil MacFarquhar, "A Powerful Russian Weapon: The Spread of False Stories," *New York Times*, August 28, 2016, www.nytimes.com/2016/08/29/world/europe/russia-sweden-disinformation.html.
[16]Kragh and Åsberg, p. 16.
[17]Ibid.

> profound disappointment over a secret agreement between Stockholm and Washington, providing U.S. sub-marines access to Swedish military bases in wartime. The telegram was immediately revealed as a Soviet forgery, but its content continued to circulate in the Swedish debate.[18]

Russian propagandists capitalized on this, not only spreading false rumors about the 2014–2015 submarines but explicitly connecting them to the old, false (yet still believed by many) rumors of secret military arrangements between the United States/NATO and the Swedish that might anger Russia and put Sweden on the frontlines of a (nuclear) World War III.

The Sweden-NATO host agreement was ratified in May 2016, but Russian influence operations continue in Sweden to this day. According to a journalist who allegedly worked undercover in Russia's Internet Research Agency (the ones responsible for Project Lakhta), Russian propagandists had their sights set on Sweden's 2018 national election.[19] And it only makes sense. The NATO threat that Russia perceives hasn't gone away, and if anything, Russian military pressure seems to be increasing in the Baltic and Scandinavia.[20] There's no reason to expect Russian information warfare in Sweden to let up any time soon.

## Fancy Bear and the Great Meme War of 2016

Russia is known or suspected to have conducted numerous other campaigns of information warfare online, but the one that has likely received the most detailed scrutiny is their effort to influence the 2016 U.S. presidential election. Russia conducted *at least* four influence operations around the 2016 U.S. presidential election aimed at removing Western roadblocks to Russian geopolitical aims, in particular the removal of Obama-era sanctions against Russia and Russian oligarchs. All of these operations—and likely others of which the American public is not (yet) aware—are directed from the Presidential Administration: Vladimir Vladimirovich Putin and his closest associates.

---

[18]Ibid., p. 9.

[19]"Journalist who infiltrated Putin's troll factory warns of Russian propaganda in the upcoming Swedish election - 'We were forced to create fake facts and news'," Jill Bederoff, *Business Insider*, published April 7, 2018, `https://nordic.businessinsider.com/journalist-who-infiltrated-putins-troll-factory-warns-of-russian-propaganda-in-the-upcoming-swedish-election---we-were-forced-to-create-fake-facts-and-news--/`.

[20]"Russia's growing threat to north Europe," *The Economist*, October 6, 2018, `www.economist.com/europe/2018/10/06/russias-growing-threat-to-north-europe`.

The first influence operation was the work of a team of hackers within Russian military intelligence (the GRU), a team known as APT28, or Fancy Bear. Fancy Bear hacked key Democrat targets and made public compromising material to discredit them and hurt Hillary Clinton's chances of winning the election. The second public influence operation was undertaken by the Internet Research Agency (IRA) on social media platforms like Facebook, Instagram, Twitter, YouTube, Tumblr, Medium, and others. They created and amplified media that supported Donald Trump, denigrated Hillary Clinton, and encouraged many groups on the American political left to vote third party or disengage from the electoral process. Both of these operations were aimed directly at American citizens, and took place in public, online.

Two other more covert and more business-oriented operations were conducted simultaneously. One involved the cultivation of human assets by building relationships between Russian operatives like Maria Butina and American public figures and business leaders, mostly (but not exclusively) on the political right. In the case of Maria Butina, who was convicted for conspiracy against the United States in 2018, the goal appears to have been to build solidarity between Russia and American conservatives, possibly in the hopes of convincing them to eliminate sanctions against Russia if/when they came to power.[21] The other operation saw Russian oligarchs and business leaders seeking to cultivate financial relationships with Donald Trump, his family, and his close associates and advisors. This is hardly different from how Putin's Presidential Administration relates to Russia's oligarchs and business leaders. Russia operates under a quid pro quo relationship between the government and businesses, enriching those involved in this relationship, and making it difficult for their competitors to do business, at least on a level playing field. Where in the United States a corrupt individual may (attempt to) bribe a public official with a payout in return for favorable legislation, in Russia it is more common to enter into a business relationship, or joint ownership of an asset, where the government official enacts favorable legislation or regulation so that the business thrives, enriching both the corrupt business person and the corrupt public official. (For a deep dive into these kinds of practices, see *Putin's Kleptocracy: Who Owns Russia?*[22])

These human relationships are of the utmost importance, as they are the end goal of the public influence operations. As significant as a close business tie with a real estate mogul and reality TV star is, that business tie or personal relationship becomes exponentially more valuable if that asset becomes the president of the United States. Not only is that asset a "friend in high places," but that asset also has a personal financial motivation to work in the interests

[21]"Maria Butina's Defiant Plea and Yet Another Russian Ploy," Natasha Bertrand, *The Atlantic*, December 13, 2018, www.theatlantic.com/politics/archive/2018/12/maria-butina-pleads-guilty-russian-agent/578146/.
[22]Karen Dawisha, *Putin's Kleptocracy* (New York: Simon & Schuster, 2014).

of his foreign business associates. Even more so should those associates, or others with whom they are coordinating, come to possess *kompromat* (the intelligence industry's term for compromising material) about the would-be president or his inner circle. As tantalizing as the investigation into these alleged and attempted relationships can be, they are ultimately not what this book is about.

This book is about how big data and information can influence people and alter history. And so, I will focus on the activity of the GRU and the IRA, two organizations that were successful at influencing public discourse, spreading falsehood, and setting the lion's share of the agenda in the final weeks of the presidential election. We may never know how many votes—popular or in the Electoral College—were won or lost by their activities. But there is no doubt that their activity influenced the issues that Americans took with them into the voting booth, and even at the end of 2018, the content they published in 2016 is still circulating and driving political conversations. (And that's to say nothing of the work they have done since then, both detected and undetected.)

Let's start with Fancy Bear.

## Fancy Bear Crashes the Democratic Party

The GRU has likely been focused on the United States for some time. They have certainly had their eyes fixed on the West in recent years. For example, in April 2015, Fancy Bear hacked into TV5Monde in France and posted messages on the national television network like, "Soldiers of France, stay away from the Islamic State! You have the chance to save your families, take advantage of it. The CyberCaliphate continues its cyberjihad against the enemies of Islamic State." While initial signs pointed to a possible ISIS attack, security experts ultimately determined the source of the attack to be an IP address associated with the GRU.[23]

One of the earliest direct signs that they intended to conduct an operation to influence or discredit the outcome of the 2016 U.S. presidential election was in October 2015, when the GRU began targeting the email account of Democratic National Committee (DNC) director of voter protection, Pratt Wiley. According to the *Chicago Tribune*, Fancy Bear hackers "tried to pry open his inbox as many as 15 times over six months."[24] This, of course, raised

[23]Sheera Frenkel, "Meet Fancy Bear, The Russian Group Hacking The U.S. Election," *BuzzFeed News*, published October 15, 2016, www.buzzfeednews.com/article/sheerafrenkel/meet-fancy-bear-the-russian-group-hacking-the-us-election.
[24]Raphael Satter, "Inside story: How Russians hacked the Democrats' emails," *Associated Press*, published November 4, 2017, www.apnews.com/dea73efc01594839957c3c9a6c962b8a.

the specter of possible hacks into voting machines to change vote counts in precincts with fully electronic voting.

While cybersecurity experts generally agree that Russia did not change any electronic vote tallies in the United States in 2016, they *did* manage to break in and steal personal information about U.S. citizens from voter registration records in states such as Illinois.[25] Election officials across the United States shored up election security and in some cases returned to paper ballots, at least as a backup. But even as the U.S. midterm election approached in November 2018, security experts and election officials across the United States were worried that many precincts around the country would not be able to withstand an attack on vote tallies if it *did* come. But even that isn't the greatest fear of many election officials. As NPR reported in a story about the Illinois data breach:

> Illinois is investing a few million dollars in federal money to help some of the state's smaller voting jurisdictions secure their data and equipment. Some counties don't even have their own IT staff. But officials here are less worried about hardware and software. They're more concerned that even a modest breach could undermine voter confidence in the machinery of our elections.[26]

Where changing voter tallies or bringing down machines on election day could be a devastating attack on the U.S. democratic system, perhaps just as effective—and certainly more *cost*-effective—would be a *discrediting* of the election results. If voter confidence were brought down far enough, it could depress turnout. And you can't recount votes that were never cast. If you targeted that voter confidence to a specific demographic or geographic region that leaned heavily one way, you could depress votes for a single party predictably and irreversibly. And if that took place in a large swing state with a large pool of electoral votes and a close final tally, you *could* theoretically swing an election.

Swinging an election isn't the only end game. Discrediting the process can call the results into question, and if enough citizens feel the election was illegitimate, it can make it difficult for the election winners to govern. This appears to be the underlying motivation of Fancy Bear's wildly successful campaign to hack the Clinton campaign and the Democratic National Committee.

---

[25]"What Illinois Has Learned About Election Security Since 2016," *All Things Considered*, National Public Radio, broadcast September 17, 2018, www.npr.org/2018/09/17/648849074/what-illinois-has-learned-about-election-security-since-2016.
[26]Ibid.

## How Fancy Bear Got In

Have you ever seen the movie *WarGames* (1983)? It's not exactly recent, but it illustrates perfectly the kind of hacker mindset that underlies the specific techniques used by Fancy Bear leading up to the 2016 election. In *WarGames*, the protagonist, David Lightman, wants to change his high school biology grade to avoid going to summer school. The school's grade database is (somewhat shockingly for 1983) accessible "online"—that is, by modem access over the phone lines from a compatible computer. The database is protected by a password that changes on a regular basis, which David doesn't have, and would only last for a short time anyway. David could try a brute-force method, like algorithmically generating random passwords until he guesses the correct one, which could work with a simple enough password and a lot of time. But it would spike system login attempts over a prolonged period of time, raising the likelihood that his attack would be detected. Instead, he uses a *human intelligence* or HumInt (pronounced *hyoo-mint*) method. He (purposefully?) gets in trouble at school on a regular basis and sent to the main office for a talk with the principal. While there, he discovers where the database password is kept and regularly notes the new password as it is changed. He gets in trouble frequently enough that his presence near the location the password is written raises less red flags than multiple access attempts from the computer at his home phone number would. And so his password theft and grade change go undetected by the school.

A little later in the movie, David is searching for the servers of a video game company so he can attempt to play a new game before it is released (and, presumably, without paying the price of new computer software in the early 1980s!). As he probes phone numbers in the geographical vicinity of the video game company, he stumbles upon a few other interesting computer connections. After making a phony reservation for a trip to Paris with a female classmate he wants to impress, he discovers what turns out to be a military server containing games that teach the kinds of strategy that could be helpful to decision-makers and negotiators at the height of the Cold War. He consults two fellow computer geeks, who tell him he'll never get through the primary system security. But he could look for a "back door"—a secret method of entry placed by the original programmer(s) unknown to the current system administrators. It turns out that one of the games on the server contains a name: Falken's Maze. David does some digging at the library and discovers a developer and inventor, Steven Falken, who had a history of working with and for the government. Upon researching his life, David discovers Falken's "back door" password and gains entry to the system. (And almost starts World War III…)

Encryption and online computer security has increased exponentially since the early 1980s. Brute-force methods, without any knowledge of the system, are nearly doomed to failure, as it is far, far easier to encrypt data than it is to computationally discover the encryption "key." (And because the same kinds of computer processors are responsible for both processes, that will continue to be the case for some time.)

But technology has also increased drastically since the 1980s, in both complexity and ubiquity. What hasn't increased since then is the human brain's capacity to manage that complexity. We make mistakes. Expert developers don't just leave back doors in their code, they leave obsolete functions, unit tests, analytic trackers, dependencies on other people's code that isn't always updated when security vulnerabilities are discovered in that *other* code, overly permissible web app installation scripts that are triggered when a system is overloaded and forced to restart—or left open when a user fails to update permissions after installation. And that's to say nothing of user error like bad passwords, reused passwords, unnecessarily running the wrong app with administrator settings, and unknowingly installing and using apps that require more system access than necessary ... and leave those doors open through their own poor security settings.

Just as fictional hackers like David Lightman used HumInt and traditional research to reduce the complexity of a computational problem, real-life hackers like those of Fancy Bear use human intelligence—and capitalize on human error—as their point of entry.

Fancy Bear's weapon of choice in 2016 was email. Specifically, a type of fraudulent email called *spearphishing*. You may be familiar with *phishing*—those usually poorly written emails saying "your email inbox is full" or "your bank account is frozen," and that to fix the problem you have to click on a link in the email and provide your personal login details—including your username and password—to solve the problem. Spearphishing is like phishing, except instead of sending the email out into the wild in the hopes of catching *some* users' data (like "fishing" with a line or a net just aims at catching *some* fish), spearphishing attacks are directed at a specific user or group of users (like a spear is aimed at a *specific* fish).

These emails look less like "your bank account is frozen" and more like "Hey [your real name], it's [your boss's real name]. I'm stuck in a meeting and need you to do X for me by lunch time." There's almost always a fraudulent link to click on that will potentially give the sender of the email access to some part of your system, or request information from you that will do the same. Or it may look like a really good copy of a Google security notice sent to your Gmail account, with a button to click on to address the security concern they detected on your account. They might be after the contents of your email account, the company documents in your G Suite account, access to your camera and microphone, or they might be trying to install a keylogger on your computer that will collect every keystroke, including usernames, passwords, URLs, and the content of the messages you write. Any of these can give them access to intelligence—such as the email addresses of coworkers and clients, internal reports, financial documents, etc.—or *kompromat*—information that could compromise the integrity of you or your firm if it became public.

On March 10, 2016, once it had become clear that Hillary Clinton would become the democratic nominee for president, her campaign team first started receiving spearphishing emails from Fancy Bear. On March 19, GRU hackers began targeting the personal Gmail account of her campaign chair, John Podesta. The attacks on the campaign continued through March and April, but also branched out to other groups linked with the campaign, such as the Clinton Foundation, the Center for American Progress, ShareBlue, and others.

The campaign was successful. One of Podesta's staffers had clicked on the link in a fraudulent Google security alert sent to Podesta's personal Gmail account on March 19 and entered his username and password. Twice. In late April, word reached Trump advisor, George Papadopoulos, that Russia had obtained *kompromat* on Clinton from "thousands of emails."[27]

Around the same time, Fancy Bear targeted the Democratic National Committee. But this spearphishing attempt didn't just target emails. The fraudulent link in those emails contained the power to install malware on the computer when clicked, and malware downloaded onto DNC computers gave the GRU access to data on both those computers and the DNC servers.[28]

Russia didn't just sit on the kompromat they gained from these attacks. In April 2016, two web sites, now believed by U.S. intelligence agencies to be connected to Russia, were registered: electionleaks.com and dcleaks.com. In June, both dcleaks.com and the blog of "hacker" Guccifer 2.0 (also believed by the U.S. intelligence community to be a likely GRU operation) began to publish kompromat gained against the Democratic Party—including information about how the DNC leadership worked to protect Clinton from the challenge posed by Bernie Sanders. This led to, among other things, a suit against the DNC from Sanders supporters—and likely a number of protest votes from leftists (either no-votes or votes for a third-party candidate like Jill Stein), hurting Clinton in the election.[29]

Also in June 2016, Julian Assange announced that WikiLeaks had obtained kompromat on Hillary Clinton. (Assange denied that Russia was the source of the data.) In July, WikiLeaks began to publish that kompromat, strategically scheduling the release of that information over the course of the remainder of the election, culminating in the release of the Podesta

[27]Raphael Satter, "Inside story: How Russians hacked the Democrats' emails."
[28]Sheera Frenkel, "Meet Fancy Bear."
[29]Raphael Satter, "Inside story: How Russians hacked the Democrats' emails."

emails on October 7, 2016—the same day (and soon after) the Trump-compromising Access Hollywood tape was released.[30]

Though we'll never know the full extent of the impact of the GRU's operation, there were several fallouts that we can likely attribute to the release of the kompromat Fancy Bear obtained from the Clinton campaign and the DNC. First, the revelations about the way the DNC held back Sanders' chances at winning the nomination alienated many left-leaning voters—likely depressing votes for Clinton both from those further left (who supported Sanders in the primaries and caucuses) and those toward the middle (who may have been demotivated to vote for Clinton, in spite of distaste for Trump). This was likely exacerbated by pro-Jill-Stein and "never Clinton" campaigns conducted by Russia's Internet Research Agency (which we'll unpack shortly).

Other details contained in the Podesta emails likely depressed turnout—or at least enthusiasm—among those same potential Clinton supporters on her left and in the middle. As Jeff Stein wrote for Vox:

> None of the Podesta emails has so far actually broken any fresh scandals about the woman on track to be the next president. Instead, they've mostly revealed an underbelly of ugliness to the multiple Clinton controversies that we've already known about: the questionable relationship between the Clinton Foundation and its donors, Clinton's ease with powerful interests on Wall Street, her ties to wealthy campaign contributors.[31]

The Clinton Foundation controversies were likely particularly damaging to Clinton's potential support among centrists turned off by Trump. Many of them had voted for Republicans in the past and were dissatisfied by the GOP's nominee. But narratives of corruption around her, combined with Republican prejudices against the Clinton name, likely made it difficult for a number of moderates and dejected Republicans to vote Democrat in 2016. The ties to Wall Street and wealthy donors likely hurt Clinton's chances with former Sanders supporters, who had been calling for Clinton to release the content of the private speeches she had been paid to give to bankers and other wealthy Americans in the past. Some of that content was awkward or embarrassing, but some of it also undercut Clinton's ability to garner support among those to the left of her campaign's policy platform.

---

[30]Ibid.

[31]Jeff Stein, "What 20,000 pages of hacked WikiLeaks emails teach us about Hillary Clinton," *Vox*, published October 20, 2016, www.vox.com/policy-and-politics/2016/10/20/13308108/wikileaks-podesta-hillary-clinton.

I also think it is likely that the months-long obsession with "Clinton's emails" compounded the impact of FBI Director James Comey's announcement on October 28, 2016, that he was reopening the investigation into Clinton's use of a private email server while Secretary of State. That investigation was reopened due to evidence that surfaced during the investigation of former Congressman Anthony Weiner, whose estranged wife, Huma Abedin, was a top Clinton aide.[32] Comey was investigating the possibility that Clinton's private email server may have contained classified information, not the content of emails published by Russia or WikiLeaks, but a fast-moving news cycle replete with vague references to "Clinton's emails" likely meant that these news stories compounded each other's impact on voters.

Perhaps most significantly, though, is that the GRU's release of kompromat on Clinton and the DNC meant that *Russia got to set a significant portion of the agenda of the presidential election.* Journalists pored over leaked content looking for scoops, which they reported. Those same journalists posed questions about the content of these documents in debates. And the candidates faced questions about these documents all along the campaign trail, from both journalists and voters. Every question about this Russian kompromat—however legitimate a concern that question represented—was a question that didn't address the policy platforms set forth by the campaigns. The hours spent chasing down security concerns and prepping answers to these questions were hours not spent by the campaigns building and defending their own platform. To a nontrivial extent, Russian hackers determined the issues that American voters would devote their limited attention to, and on which they would determine how to cast their votes on November 8. My colleagues and I call this manipulation of public messaging "weaponized truth." The contents of Fancy Bear's data dumps appeared authentic to security researchers—they were not "fake news." But they were meant to manipulate public discourse and individuals' behavior. Whether or not we call that "disinformation" (I do, because disingenuous manipulation is at the core of the operation), it is certainly a component of information warfare—and because of its basis in facts, it can be an incredibly effective one.

## Project Lakhta

Of course, Russia did not just manipulate American public discourse via the release of hacked and leaked documents. Their Internet Research Agency (IRA) also conducted information warfare against the United States through an expert social media manipulation campaign. The details of the IRA's extensive operations are detailed primarily in indictments issued in 2018 by

[32]Adam Goldman and Alan Rappeport, "Emails in Anthony Weiner Inquiry Jolt Hillary Clinton's Campaign," *The New York Times*, published October 28, 2016, www.nytimes.com/2016/10/29/us/politics/fbi-hillary-clinton-email.html.

the U.S. Department of Justice against key Russian individuals and LLCs and in two reports prepared for the U.S. Senate Select Committee on Intelligence (SSCI) and released in December 2018, providing a detailed analysis of data provided to SSCI by Twitter, Facebook (including Instagram), and Google (YouTube).[33]

The picture portrayed in these documents—which is likely still incomplete—is one of large-scale, expert manipulation of public attention through a combination of "weaponized truth," partial truths, flat-out lies, and voter suppression narratives, aimed at the election of Donald Trump as president of the United States and discrediting the potential presidency of Hillary Clinton and the democratic process in general. Further, after the election, the IRA continued to attempt to manipulate and destabilize American society, even *increasing* their activity aimed at certain American communities on platforms like Instagram. And though the data currently available suggests that government and platform activities have significantly hindered the IRA's ability to wage social media–based information warfare, it is also clear that Russian groups are still attempting to manipulate public opinion and discredit their critics via U.S.-targeted online media.

While Russian influence operations go back much further, even in the United States, the now-famous operation aimed at manipulating the 2016 election ratchets up in April 2014 with the formation of the IRA's "translator project," aimed at studying U.S. social and political groups on online platforms, including YouTube, Facebook, Instagram, and Twitter. In May 2014, the strategy for this operation, known internally as "Project Lakhta," was set: "spreading distrust towards the candidates and the political system in general"[34] with the objective of interfering in the 2016 U.S. election, specifically.[35] By June 2014, IRA operatives were already conducting in-person intelligence gathering physically in the United States (with a subsequent trip in November 2014).[36] By September 2016, Project Lakhta's *monthly* budget was approximately 1.25 million dollars.[37]

According to the IRA indictment:

> Defendants and their co-conspirators, through fraud and deceit, created hundreds of social media accounts and used them to develop certain fictitious U.S. personas into "leader[s] of public opinion" in the United States.

[33]Full disclosure: I coauthored one of those reports.

[34]United States of America v. Internet Research Agency, LLC, et al., `www.justice.gov/opa/press-release/file/1035562/download`.

[35]Ibid., p. 12.

[36]Ibid., p. 13.

[37]Ibid., p. 7.

> ORGANIZATION employees, referred to as "specialists," were tasked to create social media accounts that appeared to be operated by U.S. persons. The specialists were divided into day-shift and night-shift hours and instructed to make posts in accordance with the appropriate U.S. time zone. The ORGANIZATION also circulated lists of U.S. holidays so that specialists could develop and post appropriate account activity. Specialists were instructed to write about topics germane to the United States such as U.S. foreign policy and U.S. economic issues. Specialists were directed to create "political intensity through supporting radical groups, users dissatisfied with [the] social and economic situation and oppositional social movements."[38]

This included focusing messaging around themes like immigration, Black Lives Matter and police brutality, Blue Lives Matter, religion, and regional secession, among others. These topics were guided through internal documents provided to IRA "specialists" to use as the basis of their content, and in September 2016, one internal memo stressed that "it is imperative to intensify criticizing Hillary Clinton" in advance of the November election.[39]

The internal documents made public by the Department of Justice in their indictments of key IRA officials and shell companies only provide a small window into their actual operations targeting Americans, though. To really see what they did and how their content spread, the Senate Select Committee on Intelligence commissioned two groups to analyze private data provided to the Senate by Twitter, Facebook, and Google and report their findings. Several of my colleagues and I had the honor of contributing to one of those reports. While those datasets were all missing key metadata (and, I believe, further examples of IRA and other Russian agencies' U.S.-directed propaganda), they exhibit a *massive* and *professional* operation that far exceeds the initial statements made by platform executives. It is impossible to quantify how many votes this campaign may have changed, or at least influenced, but it is impossible to deny that this operation, in conjunction with Fancy Bear's work, was a significant factor in the tone of the election, the issues that took center stage in public discourse, and the media coverage around the election. All of this together certainly influenced *some* votes and has since cast doubt on the legitimacy of the 2016 election and fear about foreign influence of previous and subsequent elections.

But just what did their operation look like? And how pervasive was it?

---

[38]Ibid., p. 14.
[39]Ibid., p. 17.

To start with, IRA influence operations around the 2016 U.S. election hit every major platform, and even some minor ones. In addition to Facebook, Instagram, Twitter, and YouTube, evidence of IRA operations has surfaced on Google Plus, Vine, Gab, Meetup, VKontakte, LiveJournal, Reddit, Tumblr, Pinterest, Medium, and even Pokémon Go.[40] That's to say nothing of the world-wide web, where the IRA (and other branches or contractors of the Russian government) have web sites, blogs, and pro-Kremlin "think-tank" journals. This network of IRA web assets was "run like a sophisticated marketing agency" with dozens of real people posting, sharing, retweeting, and commenting on each other's memes, blogs, and tweets. As my colleagues and I wrote in our report for SSCI, "it was far more than only $100,000 of Facebook ads, as originally asserted by Facebook executives. The ad engagements were a minor factor in a much broader, organically-driven influence operation."[41] The overall budget for Project Lakhta exceeded $25 million,[42] which primarily went to paying employees to create not ads but *organic* content: tweets, posts, memes, videos, events, all shared from user accounts and pages belonging to fake personas and groups carefully crafted by IRA "specialists." Overall, IRA content pre- and post-election reached an estimated 126 million Americans on Facebook, 20 million on Instagram, and 1.4 million on Twitter. This was no tiny operation.

It's also important to note that the IRA, by and large, did not operate a network of automated accounts, known as a *botnet*. IRA employees were expected to meet daily quotas of organic posts, comments, shares, and likes. These were mainly human-operated accounts that sought to "weave propaganda seamlessly into what appeared to be the nonpolitical musings of an everyday person."[43] Thus, they had employee shifts that lined up with U.S. time zones (see the DoJ indictment previously discussed) and a system of bonuses and fines that encouraged individual "specialists" to produce high-engagement content.[44] To appear even more like real Americans, IRA "specialists" played hashtag games[45] and posted a high volume of nonpolitical content.

Project Lakhta involved far more than pro-Trump and anti-Clinton messages blasted into the ether in the hopes of reaching an audience. In fact, election-related posts only accounted for 7% of IRA Facebook content, 18% of

---

[40]Renee DiResta, Kris Shaffer, Becky Ruppel, David Sullivan, Robert Matney, Ryan Fox, Jonathan Albright, and Ben Johnson, "The Tactics and Tropes of the Internet Research Agency," New Knowledge, published December 17, 2018, `https://disinformationreport.blob.core.windows.net/disinformation-report/NewKnowledge-Disinformation-Report-Whitepaper-121718.pdf`, p. 5.

[41]Ibid., p. 33.

[42]Ibid., p. 6.

[43]Adrien Chen, "The Agency," *The New York Times Magazine*, published June 2, 2015, `www.nytimes.com/2015/06/07/magazine/the-agency.html`.

[44]Ibid.

[45]"The Tactics and Tropes of the Internet Research Agency," p. 13.

Instagram content, and 6% of Twitter content.[46] Rather, the IRA created what my colleagues and I call a "media mirage"—a false, interconnected, multiplatform media landscape, targeting multiple different communities with deceptive, manipulative messaging.[47] This "mirage" included a significant portion of apolitical content, and where the content was political, it was often focused on current divisive social issues that energized (or de-energized) members of different communities, rather than specific candidates. This mirage targeted three general communities—right-leaning Americans, left-leaning Americans, and African Americans—as well as more hyper-targeted subcommunities like pro-secessionist Texans, democratic socialists, evangelical Christians, etc. And this mirage targeted them with real news, fake news, disingenuous conversation, and—likely most significant—meme warfare.

The IRA had done their homework—both online and on the ground in the United States—when it came to targeting American communities. (And they constantly retooled their messaging based on user engagement stats, just like one would expect from a digital marketing firm.) In many cases, they targeted communities with specific messages tailored for that community, which, of course, fit the Kremlin's agenda. For example, they targeted right-wing Americans with narratives that would get them energized to come out and vote against democratic or moderate candidates—fearmongering narratives about immigration and gun rights, inspirational Christian-themed narratives, and warnings about Clinton's alleged corruption. They targeted left-leaning Americans with narratives that would de-energize them, turning their Clinton support lukewarm or encouraging a third-party or write-in protest vote—narratives about Clinton's alleged corruption, the DNC's undemocratic primary process that denied Bernie Sanders a fair shot, feminist and intersectional narratives that labeled Clinton a bad feminist. And they targeted African Americans with even more poignant voter-suppression narratives about police brutality or the racist tendencies of both parties that were intended to turn them off from the democratic process in general. The ultimate goal was to encourage votes for Trump and—if not possible to flip a leftist to vote Trump—to depress turnout from Democrats and from demographics that tended to vote Democrat. And they pursued this goal more through general social and political narratives than through posts that referenced the candidates specifically.

Notice, though, that most of these narratives are about distrust and fear—not positive American virtues that happened to be consistent with Russian aims. Yes, there were some high-engagement patriotic accounts and pro-Christianity accounts, but overall, even these accounts were about creating an insider/outsider framework that could be operationalized in other

[46]Ibid., p. 76.
[47]Ibid., p. 45.

contexts with narratives of fear or anger about outsiders. Many of the pro-Christianity accounts, for example, would frame the election as a spiritual battle between Jesus and a memeified, demonic depiction of Hillary Clinton.[48] Pro-patriotism accounts would lay the groundwork for anti-immigration, anti-Muslim, anti-LGBTQ, or anti-anything-that-could-be-construed-as-contrary-to-"traditional"-American-values content. This negativity was part of a secondary goal: to sow division, distrust, and general social discord in American society. Regardless of who wins the election, if Americans are more divided and more focused on divisive domestic issues, they are less of a partner to NATO allies and less of a threat to the Kremlin's geopolitical aims.

This secondary goal is why U.S.-directed IRA operations didn't cease in November 2016. In fact, even as the U.S. government and public started to become aware of Russian influence operations after the fact in early 2017, and account takedowns started to occur on Twitter and Facebook, *IRA Instagram activity aimed at U.S. audiences increased.* To be sure, influence operations continued on all major platforms, but the data provided by the social media companies tell a clear story of *increased* activity on Instagram in particular in 2017, especially content stoking social fears among African Americans and the anti-Trump "resistance."[49] IRA activity in 2017 targeted right-leaning audiences with predominately anti-Clinton messaging, solidifying support for Trump, and narratives about platform censorship of conservatives (as Russian accounts began to be taken down); they targeted left-leaning audiences with anti-Electoral College narratives, stoking dissatisfaction with the electoral process; and they targeted everyone with anti-Robert Mueller narratives, attempting to discredit the investigation of Russian interference in the election and potential collusion with the Trump campaign.

## What Now?

As we conclude in our Senate report, "it appears likely that the United States will continue to face Russian interference for the foreseeable future."[50] In the fallout from the 2016 election, many of the IRA's best U.S.-directed assets have been taken down and many of their key leadership and shell corporations have been indicted. This makes it far more difficult for them to conduct a high-impact operation in the United States. But not impossible. My colleagues

[48]For examples of the most popular Jesus-vs.-Hillary memes, see Kate Shellnutt, "Russia's Fake Facebook Ads Targeted Christians," *Christianity Today*, published November 3, 2017, `www.christianitytoday.com/news/2017/november/russia-fake-facebook-election-ads-targeted-christian-voters.html`. Some of the memes are also displayed and discussed in the New Knowledge report to SSCI.

[49]"The Tactics and Tropes of the Internet Research Agency," p. 93.

[50]Ibid., p. 99.

and I have continued to observe likely Russian propaganda on social media, including content directed at the 2018 U.S. midterms.[51] It's not clear yet to what extent they may have been effective, but they have not disappeared. In addition to online propaganda, we also know that Claire McCaskill's (unsuccessful) reelection campaign to the U.S. Senate was the target of hacking, likely by the GRU.[52] Russia's state-run, English-language mass media outlets, like RT and Sputnik, and hundreds of Kremlin-linked and Kremlin-aligned blogs and "journals" abound on the English-language web. We also know that Russia's geopolitical ally, Iran, has been identified as the source of online disinformation operations, which so far have received less direct scrutiny from researchers than Russia's operations. And we haven't even discussed Russia's suspected involvement in the UK Brexit referendum of 2016 or the hack-and-dump operation targeting Emmanuel Macron's presidential campaign in France in 2017.

At least for now, this is the new normal.

# Summary

In this chapter, we took a detailed look at Russian disinformation operations on social media throughout the old battlefield of the Cold War. Multiplatform influence operations in Ukraine preceded and supported military operations in Crimea and eastern Ukraine. Fearmongering narratives on television and online sought to discourage cooperation between Sweden (and Finland) and their NATO friends that might threaten the Russian energy industry. And operations from Russian military intelligence (the GRU) and a Russian contractor (the IRA) sought to destabilize and discredit the U.S. democratic process in 2016 and elect Donald Trump as president of the United States.

These operations preyed upon some of the West's greatest virtues: freedom of speech, freedom of the press, openness, and technological advances. But they also preyed upon fundamental human weaknesses: cognitive limits that make it difficult to sort fact from fiction online, tribalism, and fear of the "other." These weaknesses, combined with the reality that social evolution and legislation will always trail behind technical innovation, mean that we're likely going to be dealing with the problem of online disinformation for a long time to come.

---

[51] Jonathon Morgan and Ryan Fox, "Russians Meddling in the Midterms? Here's the Data," *New York Times*, published November 6, 2018, `www.nytimes.com/2018/11/06/opinion/midterm-elections-russia.html`.

[52] Kevin Poulsen and Andrew Desiderio, "Russian Hackers' New Target: a Vulnerable Democratic Senator," *Daily Beast*, published July 26, 2018, `www.thedailybeast.com/russian-hackers-new-target-a-vulnerable-democratic-senator`.

But this isn't just a problem for NATO countries and the former Soviet republics. This is a global problem, as both foreign and domestic groups seek to use online media to manipulate others, to their own advantage. And in countries where not only is the technology new, but *democracy* is new, online disinformation poses an even greater threat to free speech and the free flow of information, both of which are vital for a democracy to thrive.

That's the subject of the next chapter.

CHAPTER

6

# Democracy Hacked, Part 2

## Rumors, Bots, and Genocide in the Global South

New technology is neither inherently good, nor inherently bad, nor inherently neutral. When it comes to new ways to communicate and to share information, new technology irreversibly alters the social structure of a community, making new things possible and rendering the old ways of doing things inaccessible. When a community is already experiencing social change or tension, this only adds fuel to the fire. New access to information can lead to positive social change, but tools like social media can also be turned into a weapon. New tech plus new social structures can bring great instability and uncertainty to a society. That story—far more than foreign meddling—has played out repeatedly throughout the Global South.


K. Shaffer, *Data versus Democracy*,
https://doi.org/10.1007/978-1-4842-4540-8_6

## A Digital Revolution

> Technologies alter our ability to preserve and circulate ideas and stories, the ways in which we connect and converse, the people with whom we can interact, the things that we can see, and the structures of power that oversee the means of contact.[1]

These words from Zeynep Tufekci frame her discussion of how digital platforms, like social media, transform and interact with the kinds of societal and political structures in which we live and work. Think about it this way: before the invention of the automobile, many Western communities were smaller, and we lived much closer to where we worked, where we worshiped, where our children were educated, even where our food was grown. Now that we have cars, trucks, buses, trains, and airplanes, we don't need to live so close to those things. In some cases, people live more spread out; in others, people still live and work close together, but the farms that sustain those communities are far away. My parents and grandparents all grew up—and stayed—in the same county; my parents each went to the same high school as their parents did, in towns right next to each other; and when I was in high school (one more town over), I competed in track meets against my parents' and grandparents' alma maters. But when I went to college, I went out of state, and this Midwestern boy met and married a New England girl, and we've since lived in just about every part of the United States. Our kids weren't even all born in the same state.

Just as the introduction of industrial technology, and the infrastructure that supports it, gradually led to significant changes in social structures and human relationships,[2] the introduction of digital technology has done the same, only faster. For people like me, those changes have meant that I can stay connected to family, even when we live over a thousand miles away. It meant that I could pursue a career that required multiple cross-country moves. It meant that I could pursue professional development online that would allow me to make a career change without going back to (and paying for) school. And it means that my spouse could pursue her career in person in one state while I work remotely for a company in another state. The digital revolution has been a major paradigm shift for me and my family.

But for many, the revolution isn't digital.

---

[1]Zeynep Tufekci, *Twitter and Tear Gas: The Power and Fragility of Networked Protest* (New Haven: Yale University Press, 2017), p. 5.
[2]See Chapter 1 for a discussion of how this phenomenon played out at several points in human history.

The quote from Tufekci about technology's power to change social structures was written largely in reference to the Arab Spring—the wave of protests across North Africa and the Middle East that led to the overthrow of a number of autocratic governments in 2011. After many failed attempts at major governmental change in Tunisia, Egypt, and elsewhere, revolution swept much of the Arab world in 2011. What was different in 2011?

Before the Arab Spring, it was easy for people in those countries to live in a state of "pluralistic ignorance"—when you feel alone in your beliefs because you simply don't know that many others agree with you. When a government controls mass media, and people are afraid to speak out in front of all but a select few trustworthy friends and family, it is easy to be ignorant of the millions of others who share your dissatisfaction with the status quo. But digital technology, especially social media, changed that. As Tufekci writes, "Thanks to a Facebook page, perhaps for the first time in history, an internet user could click yes on an electronic invitation to a revolution."[3]

That's exactly what happened in Egypt in 2011.

In 2010, a man named Khaled Said filmed Egyptian police allegedly "sharing the spoils of a drug bust."[4] He shared this video widely, and on June 6, he died in police custody. Many believe his death to be police retaliation for his exposing their alleged corruption. On July 19, an activist created a Facebook page[5] to draw attention to Said's plight. That page was named "We Are All Khaled Said," "because all of us might face the same destiny at any point in time," according to the activist who created the page.[6]

Fast-forward to January 2011. January 25 was a holiday in Egypt, technically in honor of the police, but frequently used as an occasion for anti-government protest.[7] *We Are All Khaled Said* posted an invitation to an event—a protest in Cairo's Tahrir Square (literally translated "Liberation" Square). The invitation went viral, and ultimately hundreds of thousands of Egyptians joined an 18-day-long protest. Initially, Mubarak's government dismissed the online "clicktivism." But just like in Tunisia, the online activity of the previous six months had stirred the pot just enough to get people to come out. And when they did, their "pluralistic ignorance" was gone. Between Facebook and Tahrir Square, Egyptians desiring change realized they were not alone. They called for the ouster of Mubarak. When Mubarak realized he couldn't ignore the

---

[3] *Twitter and Tear Gas*, p. 27.
[4] Lara Logan, "The Deadly Beating that Sparked Egypt Revolution," CBS News, published February 2, 2011, www.cbsnews.com/news/the-deadly-beating-that-sparked-egypt-revolution/.
[5] "We Are All Khaled Said," Facebook page, www.facebook.com/elshaheeed.co.uk.
[6] Lara Logan, "The Deadly Beating that Sparked Egypt Revolution."
[7] Twitter and Tear Gas, p. 23.

protesters, nor could he placate them, nor imprison or murder all of them, he relented. Eighteen days after the start of the protest, he relinquished power to the military.[8]

Digital connectivity—Facebook, in particular—changed Egypt forever. Like in Tunisia, anti-government citizens in Egypt were able to discover each other, recruit others, and organize a rally that brought down the government. But that wasn't the end of things. There weren't instant elections. Rather, the military stepped in to keep order until elections could be held. And when they were held, the activists that brought down the government were not sufficiently organized into a political party or even a meaningful voting bloc. The Muslim Brotherhood stepped into the power vacuum left by the Tahrir protests, winning elections in 2012,[9] quickly attempting to move Egypt toward a more right-leaning, Islamic government. This was opposed by many in Egypt who had participated in the January 25 revolution and ultimately led to a coup[10] and early elections, which brought current Egyptian president el-Sisi into power.[11]

The turmoil that followed the January 25 revolution made one thing abundantly clear: digital connectivity is not an inevitably democratic force. Though it can transform social structures, it does not necessarily lead to equal participation, equal representation, and transparent governance. And it does not transform every society in the same way. In the case of most Arab Spring countries, digital connectivity and social networks served as an accelerant. It brought existing grievances to the surface, brought would-be revolutionaries together, and—unimpeded by existing governmental leaders who did not understand its power—upended social structures so fast that activists could not keep up. Facebook's "frictionless design"[12] made revolution so easy that no one slowed down and did the necessary work of community organizing, political party building, and planning for the future.

---

[8]"Timeline: Egypt's Revolution," *Al Jazeera*, published February 14, 2011, www.aljazeera.com/news/middleeast/2011/01/201112515334871490.html.

[9]Abdel-Rahman Hussein and Julian Borger, "Muslim Brotherhood's Mohamed Morsi declared president of Egypt," *The Guardian*, published June 24, 2012, www.theguardian.com/world/2012/jun/24/muslim-brotherhood-egypt-president-mohamed-morsi.

[10]David D. Kirkpatrick, "Army Ousts Egypt's President; Morsi Is Taken Into Military Custody," *The New York Times*, published July 3, 2013, www.nytimes.com/2013/07/04/world/middleeast/egypt.html.

[11]"Egypt election: Sisi secures landslide win," BBC News, published May 29, 2014, www.bbc.com/news/world-middle-east-27614776.

[12]Kevin Roose, "Is Tech Too Easy to Use?," The Shift, *The New York Times*, published December 12, 2018, www.nytimes.com/2018/12/12/technology/tech-friction-frictionless.html.

This is a pattern we see repeatedly throughout the world, but perhaps more starkly in the Global South. Whether Egypt, Brazil, or Myanmar, young democracies and new technology can be a volatile mix. Hand someone experiencing social upheaval and political or ethnic conflict both freedom of speech *and* a smartphone at the same time, and the results will be unpredictable at best. That's true *even when everyone is acting with the best of intentions*. But when they're not, the results can be disastrous.

Ultimately, social media and digital connectivity are a mixed bag, even when foreign state actors aren't involved. New technology is neither inherently good nor inherently bad, *nor is it inherently neutral*. Every technology has certain *affordances* and certain *limitations*, and though they may not be value-laden, they do tend toward certain kinds of social effects. When a society is already experiencing social conflict or upheaval, those effects often become more drastic and less predictable. For many countries experiencing social change and rapid technological adoption at roughly the same time, that tech has, for better or for worse (or a little of both), changed those societies forever.

## Bots in Brazil

Brazil's political landscape is fraught, fractured, and unstable. It is a young democracy, born out of dictatorship only in 1985. As of 2014, there were 28 political parties with representation in Brazil's two chambers of Congress, and governance almost exclusively happens by coalition. This can lead to uncomfortable, even disastrous alliances, like the one that reelected Dilma Rousseff to the presidency in 2014. As part of her coalition building in 2010, she named a member of another party, Michel Temer, to be her vice presidential running mate. However, when calls for Rousseff's impeachment came almost immediately after her reelection in 2014, Temer joined the opposition. When Rousseff was removed from office in 2016, Temer became the new president. While there were certainly real issues about Rousseff's presidency raised by real Brazilian citizens, it appears that social media bots played a significant role in swaying public sentiment and contributing to the calls for impeachment that were ultimately successful.

Social media–based disinformation is a staple of Brazilian politics. Political campaigns paying for the propagation of information on social media are illegal in Brazil. Brazilian law also dictates that all social media campaign activity must be controlled by "natural persons"—that is, not run via automation.[13]

---

[13]Dan Arnaudo, "Brazil: Political Bot Intervention During Pivotal Events," in *Computational Propaganda*, ed. Samuel Woolley and Philip N. Howard (Oxford: Oxford University Press, 2018), p. 136.

However, as these laws are difficult-to-impossible to enforce, automation and online rumors are part and parcel to many political campaigns in Brazil from the presidency down to local elections.

In the case of Dilma Rousseff, it appears that both her campaign and that of her 2014 presidential election opponent, Aécio Neves, used social media bots to spread campaign messages and/or smear each other. According to researcher Dan Arnaudo, Neves used bots more extensively in the presidential campaign than Rousseff.[14] Once Rousseff won the election, though, most of the anti-Rousseff bots stayed online and almost immediately joined the pro-impeachment campaign. In Arnaudo's words, "The online electoral campaign never ended, and these networks became key tools for generating support for impeachment."[15] Adding insult to injury, the fact that Rousseff's social media apparatus now belonged to the administration meant tighter restrictions and stricter oversight, giving the pro-impeachment networks a distinct advantage online.

Bots can have a variety of effects. Sometimes those effects are minimal, even null. After all, without an audience, even the most active and inflammatory bots are just "shouting" into the proverbial wind. This is particularly the case now in 2018, as platforms have taken measures to prevent (if not eliminate) automation-fueled abuse. For instance, Twitter is now more aggressive about deleting bots and preventing new accounts from appearing in search results or driving traffic toward "trending" topics until they demonstrate signs of "organic" use. Facebook has drastically pulled back on what can be done via their application programming interface (API), making automation far more difficult to accomplish. However, in 2014, the doors to the platforms were far more open, and automated posts, comments, retweets/shares, and likes/favorites were easy to manage at scale on both platforms, leading to the "hacking" of algorithmic content selection in Facebook's news feed and Twitter's real-time "trends." In fact, as Arnaudo states, on Twitter "the most retweeted messages [around the Rousseff impeachment campaign] were generated by bots."[16]

It is impossible to quantify exactly how much impact automation and other inauthentic/fraudulent online activities had on Brazil's election and impeachment results. However, it is clear from both the engagement levels of the most shared content and from the continued financial investment of campaigns—despite its illegality—that the impact is not trivial. And when

---

[14]Ibid.

[15]Ibid., p. 137.

[16]Ibid., p. 140, based on Éric Tadeu Camacho de Oliveira, Fabricio Olivetti De França, Denise Goya, and Claudio Luis de Camargo Penteado, "The Influence of Retweeting Robots during Brazilian Protests," paper presented at the 2016 49th Hawaii International Conference on System Sciences (HICSS), Koloa, DOI: 10.1109/HICSS.2016.260.

Twitter trends, Facebook news recommendations, and YouTube's "Up Next" autoplay videos significantly impact the content that users encounter, even small direct impact can be compounded when manipulated output becomes input for subsequent algorithmic analysis. (Think back to Chapter 3's discussion of the algorithmic feedback loop.)

## "Weaponizing" Facebook in the Philippines

Brazil isn't the only country to see a social media disinformation apparatus continue on, retooled, after the election for which it was first created. In May 2016, after the election of President Rodrigo Duterte, the citizens of the Philippines saw his campaign's powerful social media propaganda outfit become a critical part of the new president's administration.

Rodrigo Duterte started off at a disadvantage. If campaign spending on traditional media is any indication, he had less funds to work with than some of his rivals. However, he had a powerful social media apparatus. Duterte's team was driven online because "limited campaign funds forced them to be creative in using the social media space."[17] They took advantage of "training sessions" that Facebook provided to the presidential candidates ahead of the election. In a country with more smartphones than people and where 97% of the people who are online are on Facebook,[18] that Facebook mastery proved indispensable.

Headed by Nic Gabunada, Duterte's social media team consisted of several hundred connected volunteers. "It was a decentralized campaign: each group created its own content, but the campaign narrative and key daily messages were centrally determined and cascaded for execution."[19] (This is a similar playbook to the Russian operations in the United States that same year that we explored in the previous chapter.) Notably, the content being advanced by Duterte's supporters were not simply policy positions and innocuous statements of support. Many Duterte supporters called for violence against his critics, leading to situations like an online mob in March 2016 threatening violence, even death, against a small group of students opposed to Duterte.[20]

---

[17]Chay F. Hofileña, "Fake accounts, manufactured reality on social media," *Rappler*, last updated January 28, 2018, `www.rappler.com/newsbreak/investigative/148347-fake-accounts-manufactured-reality-social-media`.

[18]Lauren Etter, "What Happens When the Government Uses Facebook as a Weapon?," *Bloomberg Businessweek*, published December 7, 2017, `www.bloomberg.com/news/features/2017-12-07/how-rodrigo-duterte-turned-facebook-into-a-weapon-with-a-little-help-from-facebook`.

[19]Maria A. Ressa, "Propaganda war: Weaponizing the internet," *Rappler*, last updated October 3, 2016, `www.rappler.com/nation/148007-propaganda-war-weaponizing-internet`.

[20]Ibid.

Journalist Maria Ressa (founder of Rappler) calls this a "death by a thousand cuts" strategy to intimidate or silence critics, where a large number of users overwhelm a target with online threats.[21]

The problem only got worse after Duterte became president. Lauren Etter writes: "Since being elected in May 2016, Duterte has turned Facebook into a weapon. The same Facebook personalities who fought dirty to see Duterte win were brought inside the Malacañang Palace."[22] Gabunada told another Rappler journalist that "there was a need to continue campaigning [after the election] because they got 'only 40% of the votes' and needed more than that to allow Duterte to effectively govern."[23]

This involved far more than a president tweeting or even a team of bots amplifying his messages. It was part of a coordinated media strategy aimed at consolidating Duterte's power and silencing his opponents, whether politicians or journalists. Soon after the election, Duterte boycotted private media for two months, refusing to talk to any independent reporters and only pushing propaganda through state-controlled media. Simultaneously, pro-Duterte trolls and sockpuppets (fake accounts controlled by real humans) attacked independent journalists that criticized the government. When Ressa published her article, "Propaganda War," which unveiled many of these activities, she was immediately attacked by pro-Duterte trolls, including both death and rape threats and calls for her to leave the country. In Ressa's words, since Duterte's election, "when someone criticizes the police or government on Facebook, immediate attacks are posted."[24]

According to Etter, this isn't simply a case of the Duterte administration "hacking" the system or taking advantage of Facebook's targeted advertising features. In addition to providing training sessions to the Philippine presidential campaigns, after the election Facebook "began deepening its partnerships with the new administration."[25] This wasn't exceptional, but "what [Facebook] does for governments all over the world ... offering white-glove services to help it maximize the platform's potential and use best practices." Etter continues:

> Even as Duterte banned the independent press from covering his inauguration live from inside Rizal Ceremonial Hall, the new administration arranged for the event to be streamed on Facebook, giving Filipinos around the world insider access to pre- and post-ceremonial events as they met their new strongman.

---

[21] Ibid.
[22] Etter, "What Happens When the Government Uses Facebook as a Weapon?"
[23] Hofileña, "Fake accounts, manufactured reality on social media."
[24] Ressa, "Propaganda War."
[25] Etter, "What Happens When the Government Uses Facebook as a Weapon?"

Though Facebook has since made it more difficult to perform some aspects of the instant trolling that has plagued Duterte critics, the attacks on independent journalists continue under Duterte's rule. In November 2018, the Philippine government announced its intent to charge Ressa's media outlet, Rappler, with tax evasion. Rappler characterized the accusations as "intimidation and harassment."[26]

It doesn't look like this problem is going away any time soon.

# Consolidating Power in Myanmar

The Duterte administration is not the only government to use Facebook as a weapon against their own people. Recent investigations have also uncovered a large-scale information operation being conducted by the Myanmar military against the minority Rohingya people of Myanmar.

Violence and oppression are nothing new in Myanmar. Myanmar is a young quasi-democracy. It has long been ruled by dictators and autocrats, and neither government oppression nor untrustworthy state-controlled media are strangers in that country. There is also a long history of violence between the Buddhist majority and the Rohingya people, a Muslim minority population that lost their rights to citizenship almost overnight in 1982.[27]

For years, the Rohingya have been on the receiving end of legal discrimination from their government and violence from their compatriots who do not share their religious or ethnic identity. In addition to perpetuating religious discrimination in education, work, and legal standing, Myanmar's Buddhist population, often led by their monks, have routinely held rallies where they spew hate speech at the Rohingya. Not infrequently these rallies have even resulted in the death of some Rohingya people.

Both the scale of the violence and the specificity with which the Rohingya are targeted led global watchdog agencies to characterize the anti-Rohingya violence as an act of (attempted) genocide.[28]

---

[26]Alexandra Stevenson, "Philippines Says It Will Charge Veteran Journalist Critical of Duterte," *The New York Times*, published November 9, 2018, www.nytimes.com/2018/11/09/business/duterte-critic-rappler-charges-in-philippines.html.
[27]Krishnadev Calamur, "The Misunderstood Roots of Burma's Rohingya Crisis," *The Atlantic*, published September 25, 2017, www.theatlantic.com/international/archive/2017/09/rohingyas-burma/540513/.
[28]Timothy McLaughlin, "How Facebook's Rise Fueled Chaos and Confusion in Myanmar," Backchannel, WIRED, published July 6, 2018, www.wired.com/story/how-facebooks-rise-fueled-chaos-and-confusion-in-myanmar/.

Myanmar also has a "rich history" of propaganda. Timothy McLaughlin of *Wired* magazine writes that "Myanmar had spent decades reliant on state-run propaganda newspapers."[29] For just about the entire history of mass media technology—newspapers, radio, television, the internet—Myanmar's government has controlled the primary channels of media distribution. And so for those living in Myanmar today, they have known two main sources of information throughout the bulk of their lives: state-controlled media in a dictatorship and local rumors or gossip. According to former U.S. ambassador to Myanmar, Derek Mitchell, "Myanmar is a country run by rumors,"[30] and this assessment is shared by a Human Rights Impact Assessment commissioned by Facebook, which characterized Myanmar as a "rumor-filled society" up to around 2013.[31]

All of this was happening well before smartphones, Facebook, or the world-wide web entered the scene. Nevertheless, the arrival of digital technology threw fuel onto the fires already burning in Myanmar.

Two key things have changed in Myanmar in the last ten years: the arrival of the internet (which for most in Myanmar basically means Facebook) and the advent of (some aspects of) representative democracy.

While the military junta that had previously governed Myanmar was dissolved in 2011, it is generally understood that the first truly open elections in Myanmar took place in 2015.[32] A number of people in Myanmar already had access to the internet in 2015, but it was after this point when the government's censorship of the press and control over online speech began to diminish significantly, though not completely. As a result of digital technology expanding at the same time that long-standing limits on free speech were diminishing, the general populace of Myanmar exhibited a general lack of media and digital literacies necessary to navigate such an open information environment as the internet. And since "for the majority of Myanmar's 20 million internet-connected citizens, Facebook is the internet,"[33] this lack of media literacy was felt most poignantly on Facebook's platform.

The negative results were first seen in the general relationship between Buddhists and Muslims in Myanmar. The long-standing animosity between these groups was exacerbated by the newfound freedom of expression and access to a platform with significant reach. And with some aspects of anti-Rohingya discrimination baked into the law, the oppression of Muslims at the

[29]Ibid.
[30]Ibid.
[31]"Human Rights Impact Assessment: Facebook in Myanmar," BSR, published October, 2018, `https://fbnewsroomus.files.wordpress.com/2018/11/bsr-facebook-myanmar-hria_final.pdf`, p. 12.
[32]Ibid., p. 11.
[33]Ibid., p. 12

hands of the Buddhist majority was particularly facilitated by increased digital connectivity. As a report from the Brookings Institution wrote, "the sudden rollback of authoritarian controls and press censorship—along with the rapid expansion of internet and mobile phone penetration—opened the floodgates to a deluge of online hate speech and xenophobic nationalism directed at the Rohingya."[34]

The hate speech and the physical violence it spawned were not unknown problems, even in the early days. The first wave of social media–enhanced violence came in 2014. McLaughlin writes that "the riots wouldn't have happened without Facebook."[35] Much like the Arab Spring uprisings that brought down governments, the increased connectivity in Myanmar pushed tensions over a critical threshold, making the large-scale riots of 2014 possible. McLaughlin also writes that Facebook "had at least two direct warnings" of rising hate speech potentially leading to violence before those 2014 riots, but because the company saw Myanmar as a "connectivity opportunity," they dismissed the rising hate speech on their platforms and the notion that it could incite violence in "real life." That said, in 2013, the group Human Rights Watch *also* dismissed the idea that rising hate speech on Facebook was a significant problem. For most people, the primary human rights issue was one of increasing access to digital technology, as well as the social, educational, economic, and political benefits that such access would surely bring.

Of course, that's not what happened. Myanmar's citizens, so used to a combination of state-run propaganda and local rumor mills, did not immediately use social media to supplant state-run propaganda outlets with balanced and nuanced journalism. Rather, they largely transferred their participatory roles from the information economy they knew into this new platform—in other words, they made Facebook into a large-scale version of the local rumor mills to which they were accustomed to contributing. In a country plagued by such deep-seated social, political, and religious tension, it is no surprise—at least in hindsight—that the result was an increase in disinformation, misinformation, and hate speech that contributed to offline physical harm as well.

This dearth of digital literacy, propensity toward spreading and believing rumor, and deep-seated ethnic tension served as the backdrop for the psychological warfare that the Myanmar military would use against its own people.

---

[34]Brandon Paladino and Hunter Marston, "Facebook can't resolve conflicts in Myanmar and Sri Lanka on its own," Order from Chaos, Brookings, www.brookings.edu/blog/order-from-chaos/2018/06/27/facebook-cant-resolve-conflicts-in-myanmar-and-sri-lanka-on-its-own/.
[35]McLaughlin, "How Facebook's Rise Fueled Chaos and Confusion in Myanmar."

The rumors and the hate speech that encouraged violence against the Rohingya didn't just come from Buddhist citizens, many of them came from the Myanmar military. As Paul Mozur reported for *The New York Times*, there was a "systematic campaign on Facebook that stretched back half a decade" in which "Myanmar military personnel ... turned the social network into a tool for ethnic cleansing."[36] Hundreds of military personnel created fake accounts on Facebook and used them to surveil citizens, spread disinformation, silence critics of the government, stoke arguments between rival groups, and post "incendiary comments" about the Rohingya.

While this operation primarily took place on Facebook, it was multifaceted. In addition to incendiary text, the Myanmar military also used images, memes, and peer-to-peer messages, including digital "chain letters" on Facebook Messenger. One thing that did *not* appear prominently in this operation is automation. Like in the Philippines, there was no significant presence of bots. A literal army of trolls waged information warfare by means of digital sockpuppets in real time. The similarity to the operations perpetrated by Russia's Internet Research Agency during the 2016 U.S. elections is hardly coincidental. Though there is no evidence that Russia was involved in the Myanmar military's operation, there *is* evidence that Myanmar military officers traveled to Russia and studied their information warfare tactics.[37]

The military's primary target in this psychological warfare operation—or PsyOp—was the Rohingya people. But in 2017, some of their operations aimed at both sides, spreading disinformation to both Buddhists and Muslims, telling each that an attack was immanent from the other side. According to Mozur, "the purpose of the campaign ... was to generate widespread feelings of vulnerability and fear that could be solved only by the military's protection." The fledgling, young democracy posed a threat to the military that until very recently ran the country. They took advantage of the people's general lack of digital media literacy and their inexperience with democratic free expression, and turned that vulnerability into a weapon aimed at their own people. And by targeting the Rohingya, who were already the victims of misinformation and hate speech online, they threw fuel on an existing fire, contributing significantly to one of the largest humanitarian crises of the twenty-first century.

---

[36]Paul Mozur, "A Genocide Incited on Facebook, With Posts From Myanmar's Military," *The New York Times*, published October 15, 2018, `www.nytimes.com/2018/10/15/technology/myanmar-facebook-genocide.html`.
[37]Ibid.

# Success in the Latin American Elections of 2018

Summer 2018 was a busy time for Latin American politics. There were presidential elections in Colombia and Mexico, political theater masquerading as an election in Venezuela, and major unrest in Nicaragua, which was riddled with anti-government protests. I spent the summer primarily focused on monitoring and reporting on the elections in Colombia and Mexico and studying the Latin American (dis)information landscape more broadly. We observed nothing at the scale of what took place in the United States, the Philippines, or Myanmar—likely (hopefully!) because the social platforms have learned some lessons and started to implement changes that make it harder (though by no means impossible) to conduct effective disinformation operations. However, a counter-trend also emerged, where instead of one bad actor launching a campaign for or against a particular candidate, party, or people group, we observed many smaller operations in service of a variety of candidates and interest groups. We also observed a new trend that may explain the smaller-scale operations on Twitter and Facebook, but also poses a major threat to democracies in the near future: peer-to-peer messaging. Let's unpack these trends.

First, the social networks. As my colleagues and I summarize in our postelection report:

> We analyzed content broadly collected on Twitter via key search terms, as well as more selective targets on Facebook. Overall, we found evidence of nine coordinated networks artificially amplifying messages in favor of or against a candidate or party in Mexico, and two coordinated networks in Colombia. While the volume of coordinated information operations in Colombia was noticeably lower, one of those operations was international in focus, aimed at stoking anti-government sentiment and action in Colombia, Mexico, Nicaragua, Venezuela, and Catalonia. While some of these networks were difficult to attribute, we traced some back to the responsible person(a)s, and at least two of them show indications of foreign involvement.[38]

Most of these networks were automated networks with little sophistication and, as far as we could tell, little impact. In fact, one of them, which was only online for roughly 48 hours, demonstrates the progress that the social platforms have made in detecting and removing *botnets*. This was a network of 38 Twitter accounts, controlled by an individual we traced on

[38] "2018 Post-Election Report: Mexico and Colombia," New Knowledge, accessed December 1, 2018, www.newknowledge.com/documents/LatinAmericaElectionReport.pdf.

Facebook, each posting hundreds of messages per day that were taken from a small library of posts to choose from, resulting in hundreds of identical copies of each of these posts across the network. These messages were always in opposition to Ricardo Anaya, the presidential candidate who ultimately finished second, or in support of long-shot candidate Jaime Rodríguez Calderón, nicknamed "El Bronco." We dubbed this botnet the Bronco Bots.

The Bronco Bots were very short-lived. We discovered them almost immediately and reported them to Twitter, and within hours of our report—and just two days after they came online—they were suspended. It may or may not have been our report that resulted in the suspensions. In fact, Twitter has gotten much better since 2016 at identifying mass automation and suspending the accounts independent of any reports from users or researchers. That's obviously a good thing. But we discovered several networks similar to the Bronco Bots, supporting a variety of candidates, and only some of them were suspended by the platform before the election (if at all). It's also worth noting that in the course of an event unfolding in real time—like the Unite the Right rally in August 2017 or the chemical weapons attack in Syria in April 2017—a network of bots or fake accounts can do a lot of damage in less than 48 hours, especially if they are amplifying a (false) narrative being pushed by real people or by nonautomated fake accounts controlled by real people. For example, we uncovered another anti-Anaya network on Twitter using automation to amplify a set of YouTube videos advancing already debunked rumors about alleged corruption from Anaya. By targeting the evening of and the day after the final presidential debate, they were able to conduct their operation without worry of account suspensions interfering until attention had already organically dwindled.

A more insidious, and longer-lasting, botnet arose to stoke anti-government sentiment, and even promote violence, across Latin America and elsewhere. The accounts in this network were generally created a few days or weeks before operations began and primarily pushed anti-government messages to targeted audiences in Venezuela, Colombia, and Nicaragua, with links to content on YouTube and two recently created web sites.

The account profiles were all variations on a theme, and locations given in the account profiles appeared to be fake (in some cases, the city-country combinations they claim as home simply do not exist). Though these accounts claimed to be different individuals in different countries, the content posted to these accounts was often identical and always fast-paced and high-volume. In addition to anti-government messaging in Venezuela and Nicaragua, these accounts actively attempted to link Colombian presidential candidate, Gustavo Petro, to the FARC terrorist organization. FARC and Colombia were at war until an unpopular peace deal was signed in 2016. Petro's support for the peace deal, combined with his leftist politics, left him open to characterizations of being a communist or a terrorist, which this botnet seized upon.

(Petro ultimately lost the runoff election to Iván Duque, a right-wing populist and critic of the Colombia-FARC peace agreement.) While most posts from this botnet were in Spanish, a few English tweets slid through now and then. These included links to tech tutorials and a site that focuses on sensitive social issues in the United States.

And occasionally Twitter automation and analytics tools (oops).

It is clear that the individual or group behind this particular botnet in Latin America was taking steps to mask their location. There were also indications that they may not be from Latin America, such as the accidental posting of English-language content and the lumping together of target audiences speaking the same language, but living in different countries—even continents, with the inclusion of posts targeting users in Catalonia. This may indeed have been a foreign influence operation. But unfortunately—or is it fortunately?—the network was taken down by Twitter before we could make a high-confidence origin assessment.

The biggest threat in Latin America wasn't Twitter, though. It also wasn't Facebook, or Instagram, or YouTube. It was peer-to-peer messaging, primarily on the Facebook-owned platform, WhatsApp.

WhatsApp is a messaging service that supports text, voice, and video calling, as well as file sharing, over an encrypted data connection. For some people, it represents a less public (and less surveilled) way to connect with friends and family than Facebook or Twitter. For others it provides voice, video, and text all over wifi or a data connection, saving them money on their monthly phone bills. Regardless of the reason, WhatsApp is becoming increasingly popular in some parts of the world, particularly the Spanish-speaking world. In many places where Facebook usage is on the wane, and Twitter never really made a splash, WhatsApp is alive with digital communities and information sharing. For instance, according to Harvard's Nieman Lab, WhatsApp is the most popular social platform in Mexico.[39]

WhatsApp and other private messaging apps (like Signal, Facebook Messenger, Slack, even good-old-fashioned text messaging and email) pose a significant challenge for disinformation researchers and fact-checkers. Because of the high level of connectivity for many who use the app, it is easy for both true and false information to spread, even to become viral, on WhatsApp. But because the messages are private and encrypted, there is no easy way to see what is trending on WhatsApp and in what communities.

---

[39]Laura Hazard Owen, "WhatsApp is a black box for fake news. Verificado 2018 is making real progress fixing that.," Nieman Lab, published June 1, 2018, `www.niemanlab.org/2018/06/whatsapp-is-a-black-box-for-fake-news-verificado-2018-is-making-real-progress-fixing-that/`.

As we've already explored, psychologists and rumor experts DiFonzo and Borgia identify four primary factors that contribute to whether or not someone believes a claim that they encounter:

- The claim agrees with that person's existing attitudes (confirmation bias).
- The claim comes from a credible source (which on social media often means the person who shared it, not the actual origin of the claim).
- The claim has been encountered repeatedly (which contributes to *perceptual fluency*).
- The claim is not accompanied by a rebuttal.[40]

Perhaps the biggest obstacle to countering disinformation and misinformation on private chat apps like WhatApp is that we don't know what claims are encountered repeatedly (going viral), what the sources of those claims are, or what audiences they are reaching (and what biases they already hold), and so fact-checkers don't know what needs rebutting. Because of these obstacles, when I talk to researchers and policymakers concerned with Latin America, peer-to-peer messaging is their biggest fear. And having seen the anti-surveillance writing on the wall, it's a growing fear among researchers and policymakers in countries where Facebook, Instagram, and Twitter still dominate, too.[41]

But all hope is not lost. One initiative that took place during the 2018 Mexican election proved that it is possible to expose and rebut rumors and misinformation on private communication apps like WhatsApp. That initiative was called Verificado 2018.

Verificado 2018 was a collaboration between Animal Político, Al Jazeera, and Pop-Up Newsroom, supported by the Facebook Journalism Project and the Google News Initiative. Their goal was to debunk false election-related claims and to do so in ways that are in line with how users use the platforms the claims are on. Because of WhatsApp's dominance in the Mexican social media landscape, Verificado focused heavily on operations there.

---

[40]Prashant Bordia and Nicholas DiFonzo, Rumor Psychology: Social and Organizational Approaches (Washington, D.C.: American Psychological Association, 2017).

[41]I should note that, while private chat applications make it harder to discover disinformation operations on those applications, those applications are still indispensable to researchers. That's because we care about the privacy of our communications, and end-to-end encrypted messaging apps are key to avoiding surveillance from adversaries, especially those with the power of governments behind them. As of 2018, Signal, from Open Whisper Systems, is the most often recommended encrypted communication app from security researchers, vulnerable activists, and tech journalists.

Because WhatsApp is a peer-to-peer platform, oriented around sharing messages with individuals or small groups of friends and family, Verificado took a small-scale, social approach to fact-checking. They set up a WhatsApp account where *individuals* could send them information in need of verification. Verificado's researchers would then respond *individually* with the results of their research. This personal touch (which one has to assume involved a fair bit of copy-and-paste when they received multiple inquiries about the same claim) was more organic to the platform and allowed users to interact with Verificado more like the way they interact with others on the platform.

Now, in the face of a coordinated disinformation campaign during the course of a massive general election that involved thousands of individual races, cut-and-paste only scales so far. So Verificado also crafted their debunks in ways that would promote widespread sharing, even virality, on WhatsApp. Multiple times a day, they updated their public status with one of the debunks that was prompted by a private message they received. These statuses could then be shared across the platform, much like tweets or public Facebook posts. They also created meme-like images that contained the false claim along with their true/false evaluation stamped on it. This promoted user engagement, associated their true/false evaluation with the original image in users' minds, and promoted viral sharing more than simple text or a link to a web article would (though, they did post longer-form debunks on their web site as well).[42]

For the same reasons that it is difficult to study the reach and impact of misinformation and disinformation on WhatsApp, it's difficult to quantify the reach and impact of Verificado 2018's work. But the consensus is that they had a nontrivial, positive impact on the information landscape during a complicated, rumor-laden election cycle, and they made more progress on the problem of private, viral mis-/disinformation than just about anyone else to date. They even won an Online Journalism Award for their collaboration.[43]

Peer-to-peer disinformation isn't going away. As users are increasingly concerned about privacy, surveillance, targeted advertising, and harassment on social media, they are often retreating to private digital communication among smaller groups of people close to them. For many, the social media honeymoon is over, the days of serendipitous global connection gone. Safety, security, and privacy are the new watchwords. In some cases, this means

[42]Owen, "WhatsApp is a black box for fake news."

[43]"AJ+ Español wins an Online Journalism Award for Verificado 2018," Al Jazeera, published September 18, 2018, https://network.aljazeera.com/pressroom/aj-español-wins-online-journalism-award-verificado-2018.

less exposure to rumors and psychological warfare. In other cases, it simply means those threats are harder to track. This is by no means a solved problem, but examples like Verificado 2018 give us hope that solutions are possible and that we might already have a few tricks up our collective sleeves that can help.

## Summary

In this chapter, we've explored recent online disinformation operations in the Global South. From Latin America to Northern Africa to Southeast Asia, we have seen the way that social media platforms have amplified rumors, mainstreamed hate speech, and served as vehicles for psychological warfare operations. In some cases, this misinformation and disinformation has fueled not only political movements and psychological distress but has also motivated offline physical violence and even fanned the flames of ethnic cleansing and genocide.

The problem of online disinformation is bigger and more diverse than many in the West realize. It's bigger than "fake news," bigger than Russia and the American alt-right, bigger than the Bannons and the Mercers of the world, bigger than Twitter bots, and even bigger than social network platforms. As long as there has been information, there has been disinformation, and no society on our planet is immune from that. This is a global problem, and a human problem, fueled—but not created—by technology. As we seek to solve the problem, we'll need a global, human—and, yes, technical—solution.

CHAPTER

7

# Conclusion

## Where Do We Go from Here?

Information abundance, the limits of human cognition, excessive data mining, and algorithmic content delivery combine to make us incredibly vulnerable to propaganda and disinformation. The problem is massive, cross-platform, and cross-community, and so is the solution. But there are things we can do—as individuals and as societies—to curb the problem of disinformation and secure our minds and communities from cognitive hackers.

## The Propaganda Problem

Over the course of this book, we've explored a number of problems that leave us vulnerable to disinformation, propaganda, and cognitive hacking.

Some of these are based in human psychology. *Confirmation bias* predisposes us to believe claims that are consistent with what we already believe and closes our mind to claims that challenge our existing worldview. *Attentional blink* makes it difficult for us to keep our critical faculties active when encountering information in a fast-paced, constantly changing media environment. *Priming* makes us vulnerable to mere repetition, especially when we're not conscious of that repetition, as repeated exposure to an idea makes it easier for our minds to process, and thus believe, that idea. All of these traits, developed over aeons of evolutionary history, make it easy for us to form biases and stereotypes that get reinforced, even amplified, over time, without any help from digital technology.


K. Shaffer, *Data versus Democracy*,
https://doi.org/10.1007/978-1-4842-4540-8_7

Some of the problems are technical. The excessive mining of personal data, combined with *collaborative filtering*, enables platforms to hyper-target users with media that encourages online "engagement," but also reinforces the biases that led to that targeting. Targeted advertising puts troves of that user data *functionally*—if not *actually*—at the fingertips of those who would use it to target audiences for financial or political gain.

Some of the problems are social. The rapidly increasing access to information and people that digital technology affords breaks us out of our *pluralistic ignorance*, but we often aren't ready to deal with the social implications of information traveling through and between communities based primarily on what sociologists call *weak ties*.

When information abundance, human psychology, data-driven user profiling, and algorithmic content recommendation combine, the result—unchecked—can be disastrous for communities. And that appears to be even more the case in communities for whom democracy and (relatively) free speech are also new concepts.

Disinformation is fundamentally a *human* problem. Yes, technology plays its part, and as argued earlier in this book, new technology is neither inherently bad nor inherently good *nor inherently neutral*. Each new technology has its own affordances and limitations that, like the human mind, make certain vulnerabilities starker than others. But ultimately, there is no purely technical solution to the problem. Disinformation is a behavior, perpetrated by people, against people, in accordance with the fundamental traits (and limitations) of human cognition and human community. The solution necessarily will be human as well.

Of course, that doesn't mean that the solution will be simple. Technology changes far more rapidly than human biology evolves, and individuals are *adopting* new technologies faster than communities are *adapting* to them. Lawmakers and regulators are probably the furthest behind, as many of the laws that govern technology today—in the United States, at least—were written before the advent of the internet.[1] Perhaps most striking, though, is the surprise that many of the inventors of these technologies experience when they witness the nefarious ways in which their inventions are put to use. If the inventors, whose imagination spawned these tools, can't envision all of the negative ends to which these technologies can be directed, what chance do users, communities, and lawmakers have?!

[1]Key U.S. laws written before the advent of the internet include the Computer Fraud and Abuse Act (1984), the Federal Educational Rights and Privacy Act (1974), and, for all practical purposes, the Health Insurance Portability and Accountability Act (1996).

The solutions may not be simple, and we may not be able to anticipate and prevent all antisocial uses of new technology, but there are certainly things we can do to make progress.

Consider the bias-amplification flow chart from Chapter 3. It is tuned primarily for search engines, but it applies to most platforms that provide users with content algorithmically. Each of these elements represents something that bad actors can exploit or hack. But they each also represent a locus of resistance for those of us seeking to counter disinformation.

For example, the Myanmar military manipulated existing social stereotypes to encourage violence against the Rohingya people and dependence on the military to preserve order in the young quasi-democratic state. Using information about those existing stereotypes, they created media that would exacerbate those existing biases, encouraging and amplifying calls to violence. They not only directly created pro-violence media and amplified existing calls to violence, they also affected the content database feeding users' Facebook timelines and created content that encouraged user engagement with the pro-violence messages. Thus, the model, which took that content and user activity history as inputs, further amplified the biased media delivered to users in their feeds. And then the cycle began again.

Similar cycles of bias amplification have led to increased political polarization in the United States, as we explored in Chapters 4 and 5. Even the perpetrators of harassment during GamerGate, many of whom we would likely consider to be radicalized already, engaged in a game of trying to outdo each other in engagement, victim reaction, or just plain "lulz." Since content that hits the emotions hardest, especially anger, tends to correlate with stronger reactions, it's not surprising that the abhorrence of GamerGate accelerated furiously at times, as GamerGaters sought to "win" the game.

Users can counter this vicious cycle, in part, by being conscious of their own personal and community biases and engaging in activity that provides less problematic inputs to the model. I *don't* mean employing "bots for good" or "Google-bombing" with positive, inspirational messages. I'm a firm believer that any manipulative behavior, even when employed with good intentions, ultimately does more social harm than social good. Rather, what I mean is using one's awareness of existing individual and community biases and making conscious choices to resist our "defaults" and the biases they represent.

For example, one of my digital storytelling students was a young woman of color who realized that when she chose visual media for her blog posts, she was choosing the "default" American image—one that was very white- and male-oriented. So she resolved in her future projects to include images that represent her own demographic, doing a small part to bring the media landscape more in line with the diversity that is actually present in our society. Remember, it's the defaults—both mental and algorithmic—that tend to reinforce existing bias. So by choosing non-defaults in our own media creation and consumption, and by being purposeful in the media we engage with, we can increase the diversity, accuracy, and even justice of the content and user activity that feeds into content recommendation models.

However, individual solutions to systemic problems can only go so far. But as Cathy O'Neil argues in her book *Weapons of Math Destruction*, the same data, models, and algorithms that are used to target victims and promote injustices can be used to proactively intervene in ways that help correct injustices.[2] Activists, platforms, and regulators can use user activity data and content databases to identify social biases—even unconscious, systemic biases—and use those realizations to trigger changes to the inputs, or even the model itself. These changes can counter the bias amplification naturally produced by the technology—and possibly even counter the bias existing in society. This is what Google did to correct the "Did the holocaust happen?" problem, as well as the oppressive images that resulted from searches for "black girls" that we discussed in Chapter 3.

Now, such corporate or governmental approaches to rectifying social injustices lead quickly to claims of censorship. Intervening in the content and algorithmic recommendations will simply bring in the bias of programmers, at the expense of the free speech of users, making the platforms the biased arbiters of truth and free expression. This is a very real concern, especially seeing how often platforms have failed when they have attempted to moderate content. But if we begin from the realization that content recommendation engines amplify existing social bias by default, and that that bias amplification itself limits the freedom of expression (and, sometimes, the freedom simply to live) of certain communities, it can give us a framework for thinking about how we can tune algorithms and policies to respect all people's rights to life, liberty, and free expression. Again, it's not a simple problem to solve, but as long as doing nothing makes it worse (and current data certainly suggests that it does), we need to constantly reimagine and reimplement the technology we rely on in our daily lives.

Take Russia's activity in 2016 as another example. In contrast to the Myanmar military, much of their activity began with community building. They shared messages that were in many ways innocuous, or at least *typical*, expressions of in-group sentiment. This allowed them to build large communities of people who "like" (in both the real-world sense and the Facebook sense) Jesus, veterans, racial equity, Texas, or the environment. The more poignant attacks and more polarizing messages often came later, once users had liked pages, followed accounts, or engaged regularly with posts created by IRA "specialists." The result was a "media mirage," and in some sense a "community mirage,"

---

[2]Cathy O'Neil, "Weapons of Math Destruction: How Big Data Increases Inequality and Threatens Democracy," (New York: Broadway Books, 2017), p. 118.

where users' activity histories told the content recommendation model to serve them a disproportional amount of content from IRA-controlled accounts or representing Kremlin-sympathetic views. This not only gave the IRA a ready-made audience for their fiercest pro-Trump, anti-Clinton, and vote-suppression messages as the election neared. It also meant that users who had engaged with IRA content were more likely to see content from *real Americans* who held views that, no matter how real and personal, happened to align with the Kremlin's interests. By reinforcing existing social biases and personal interests early on, the IRA were able to "hack" the model so that users most likely to engage Kremlin-friendly messages saw more of them, even if the sources were legitimate.

This is a situation where the contrast between disinformation as *manipulative behavior* and disinformation as *deceptive messages* comes into sharp relief. By researching and weaponizing existing social biases, the IRA was able to wield those earnestly held beliefs in their efforts to manipulate unwitting Americans into taking actions they may or may not have otherwise taken. And that was true regardless of whether or not the posts and memes were factual.

No amount of individual fact-checking will solve this problem. This is first and foremost a *behavioral* issue that can only be seen and solved with the analysis of data on the large scale: cross-community, cross-platform, over long stretches of time. But if platforms, governments, and/or third-party researchers conduct that large-scale analysis, and accounts are suspended for this coordinated, inorganic, manipulative behavior, then those more sophisticated state actors lose many of their best tools for manipulating communities into unwittingly doing their bidding. And in the process, we all regain the integrity of our information space.

That integrity doesn't come easy, especially when we are facing influence operations from sophisticated actors like a military or state-sponsored firm. As Russia's operations in particular demonstrate, disinformation is international and cross-platform and weaponizes fact and fiction alike. As a result, no one entity has all of the necessary data or expertise to solve the problem. Platforms don't have each other's data, nor the data held by intelligence or law enforcement agencies. And they do not have the cultural or linguistic experts that the intelligence agencies, universities, and nonprofit think-tanks do. Governments don't have the platforms' data—and can only collect data under certain legal circumstances—nor do they have many employees with significant engineering experience. And while third-party researchers have, at times, a healthier mix of experts in different areas, they also have less direct access to platform data. Just like the U.S. government realized after the September 11 attacks, information sharing and the collaborative utilization of a variety of areas of expertise will be key to addressing the most sophisticated disinformation campaigns. And given the legal constraints Western governments rightly face when it comes to surveilling their own citizens, and the competitive relationship the tech companies have with each other, it is likely that carefully structured public-private partnerships will be required to address the disinformation defense needs that societies around the globe are increasingly encountering.

## But What Can I Do?

It's clear that disinformation is a large-scale problem that requires large-scale solutions. Platforms, governments, and researchers must all do their part to address this rapidly evolving problem collaboratively. But is there anything that individuals can do?

We've already addressed a few ways that individuals can alter the inputs that big data algorithms use to determine the content that we see on our favorite platforms. Mindfulness about what we consume, what we engage, what we share, and what we create is absolutely essential, both to fight "fake news" and to fight the amplification of social bias and polarizing messaging.

Similarly, we can limit the data about us that individual algorithms have access to. Consuming some content offline, registering for different services with different email addresses, using different browsers for different services, even paying cash and avoiding buyer rewards programs can keep our personal data in different, smaller piles rather than one big one that many platforms can access. The result may be less relevant ads (is that really a problem?!), but also less effective hyper-targeting with messages designed to manipulate.

But ultimately, some networked activism and collective political action will be necessary. Elected leaders, regulators, and platform executives need to be involved in solving the propaganda problem, and to do that they need to be motivated—electorally, legally, financially. But for that to work, we don't just need governments and corporations working together, all of us need to work together.

Democracy depends on the free flow of information, both to inform and to afford all of us the chance to deliberate and persuade each other. If we can't trust the integrity of the information we consume, we can't trust the democratic process.

I opened Chapter 6 with a quote from Zeynep Tufekci that bears repeating. "Technologies alter our ability to preserve and circulate ideas and stories, the ways in which we connect and converse, the people with whom we can interact, the things that we can see, and the structures of power that oversee the means of contact."[3] *Technologies alter the structures of power*. That's not

[3]Zeynep Tufekci, *Twitter and Tear Gas: The Power and Fragility of Networked Protest* (New Haven: Yale University Press, 2017), 5.

necessarily a bad thing (remember Ferguson), nor is it necessarily a good thing (remember GamerGate), but it's absolutely not a neutral thing. It's up to us to invent, implement, and rein in technologies in ways that bring our vision of what the world could be to fruition.

Too often we are passive, going with the flow, letting the affordances of the technology—and its manipulation by more motivated, and often less ethical, actors—determine the future we are building. If we want to make a positive mark on the universe, we need mindfulness, deliberateness, and collaboration, as we step into the next chapter of our history. The problem likely won't go away anytime soon. But that shouldn't stop us from working to minimize its effects and harness the power of new technologies for good.

# Index





K. Shaffer, *Data versus Democracy*,
https://doi.org/10.1007/978-1-4842-4540-8

## M



## N, O



## P



## Q



## R



## S

Made in the USA
Middletown, DE
30 November 2021